Eλ

for 2020 through 2024

License Examinations

Stephen Horan, NM4SH

US Amateur Radio Licensing Series

**Amateur Radio Extra Class Licensing
for 2020 through 2024 License Examinations**

Fifth printing with URL updates: April 2023

Paperback ISBN-13: 979-8616179494

Cover image by the author.

Author

Stephen Horan, NM4SH, has been a licensed amateur for over 25 years. During that time, he helped develop and lead weekend licensing classes with the Mesilla Valley Radio Club in Las Cruces, NM. Now, he mainly operates with digital modes from his home in Virginia, where he is member #1501 of the PODXS 070 Club, has the Loyal Order of Narrow-banded Phase-shifters (LONP) certificate #266, and is an ARRL-certified Volunteer Examiner.

Steve's professional experience includes 23 years as a professor of Electrical Engineering at New Mexico State University and 12 years as an engineer for NASA, including four years as an RF spectrum manager. In addition to this series of Amateur Radio licensing study guides, he is the author of the textbook *Introduction to PCM Telemetering Systems* and has published numerous technical articles in journals and conferences.

Visit his author page at `https://nm4sh.net/`. You can reach him via e-mail at `nm4sh@arrl.net`.

Also by the Author of this Amateur Radio Licensing Series:

- Amateur Radio General Class Licensing

- Amateur Radio Technician Class Licensing

Contents

List of Figures

List of Tables

BACKGROUND

Note to Readers

Welcome back for your last licensing step as you advance your knowledge of amateur radio. With the Extra Class license, you will have full access to the amateur bands in the United States (US). The amateur community will also expect you to demonstrate the highest technical and operating skills.

As with the Technician and General Class license examinations, the Extra Class examination uses standardized questions, so everyone seeking this amateur radio license in the US will see questions drawn from the same master question pool. Amateurs participating in the National Conference of Volunteer Examiner Coordinators (NCVEC) (http://www.ncvec.org) design and maintain the question pool. This version of the question pool is valid from 1 July 2020 through 30 June 2024. The NCVEC will post any changes to the question pool on their Web site. I have designed this study guide specifically for the Extra Class license examination. The administrators for your license examination will be amateur radio volunteers, usually local, certified to give the test. They have all been through this same process. You will be among friends.

This study guide enables currently-licensed amateurs to pass the top-level Extra Class examination by building on their knowledge of electronics and amateur radio practices. Because this is a study guide directed at the examination questions, there continues to be a certain level of compromise in the presentation. I assume that you have basic knowledge and operational experience as a General Class amateur, and you desire to improve your knowledge and skills. As you know through your experience as a General Class license holder, there is always more to learn about radio science, radio operations, and the Federal Communications Commission (FCC) rules. That will continue after you earn your Extra Class license.

As with the prior study guides, each chapter covers one part of the Extra Class question pool. The chapter starts with a short discussion of key concepts behind several of the questions in the chapter that need more context than the space available for the questions' responses. Then, I introduce you to the questions and answers, with the correct answer identified in bold type and the explanation for why it is correct. In many cases, the answer's explanation goes beyond merely picking the correct answer among the options given. To help you learn about amateur radio while we go through the questions, I also include reasons why many incorrect options are wrong. I strive to provide sufficient instruction to ensure successful completion of the Extra Class exam element and learn more advanced amateur radio techniques. Once you have successfully passed the examination and have upgraded your privileges, I encourage you to pursue the theory and practice of amateur radio more deeply and work towards being an example of a skilled operator.

"How to Study" Suggestions

Given that you are doing this preparation in addition to your other activities, I offer the following suggestions for efficiently using your time.

Become familiar with the material   Before you can study effectively, you need to know where you are going! To do this, look over the question pool and the explanations. The general format is the same on all amateur radio exam elements. There are "technical parts," "operational parts," "safety parts," and "rules and regulations parts." Get a general sense of the flow of the material and the level of detail required. For the Extra Class exam, the level of electronics and mathematics is higher, building on the Technician and General Class license knowledge. However, you still do not need an engineering degree to pass.

Assess what you already know and what is new   Based on your experience and knowledge of amateur radio as a General Class license holder, parts of the material will look familiar. Other aspects, perhaps the complex technical concepts, will be new to you. For now, try to identify those specific concepts that are new to you and place most of your energy there. Also, be sure you understand the familiar material's specific amateur radio nuances.

Prioritize what is new to you   Depending upon your background and interests, you will grasp some new concepts quickly, while others will leave you wondering what the question is asking. Sort the new topic areas by the level of difficulty for you. Give yourself some confidence by working through the concepts that seem easier to you and building towards the complex concepts.

Make a list of questions   Write down the difficult concepts for you. Consult resources to which you have access. For example, you can use amateur radio friends, Web references in the study guide, or Wikipedia entries to help you understand the question and the correct answer. If there is an amateur radio club in your area, members will be ready to help you.

Make study aids   You can use study aids to help you track the key concepts that you need to learn and help you to become familiar with the material. You can learn items, such as frequency allocations, through experience or memorization. With full amateur band access privileges exceeding your current access, you may need to memorize the new privileges. To make the memorization easier, make a set of index cards with the frequencies, power limits, and types of emissions allowed. You can also do this with the necessary equations for antenna design, modulation characteristics, or other concepts that you are trying to master. Review the index cards a few times each day until you feel more comfortable with the ideas. Do not try to memorize everything in one sitting.

Test yourself   You can take one of the online practice exams to see if you are ready for the real thing. Visit sites such as `https://www.qrz.com/hamtest/`, `http://www.arrl.org/exam-practice`, or `http://www.eham.net/exams/`. You can find more sites by using a search engine in your Web browser. When you score consistently above 75% on the practice tests, you may be ready.

Relax   Remember, you do not need a score of 100% to pass the exam! Do as well as you can in learning the concepts you can grasp. The commentary on the questions will help you understand which of the four choices for the answer is correct and which is incorrect. Learn to recognize the distraction options you can eliminate by using the comments for the question pool as you study. Do not put yourself under pressure to memorize everything. Be willing to tell yourself that you can write off a few questions that you cannot get at this time and hope for the best when you take the actual exam. Some concepts will become more understandable as you operate with your new privileges.

About the Exam

The FCC mandates that the question pool for the amateur service license examinations have a certain structure. The NCVEC has a question pool committee that designs and publishes the questions used in the exam. The three major license elements are
Element 2 — Technician Class License
Element 3 — General Class License
Element 4 — Extra Class License
Note: there no longer is an Element 1; that was the Morse code examination.
   The question pool designers then break the license elements into ten subelements. The designers divide the subelements into question groups reflecting the question pool designers' curriculum. The question groups cover topics such as radio theory, operations, safety, and FCC regulations. The Extra Class question pool contains over 600 questions. Each subelement has between one and eight groups of at least 11 questions. Each license exam for Elements 2, 3, and 4 will use one and only one question from each group regardless of the number of

questions in the question pool for that group. Use this knowledge to help you to design your studies. It would be best if you did not try to memorize the answer to each question but learned the general principles.

Table 1 shows how the designers organized the question pool for the Element 4 Extra Class license examination. The Extra exam has 50 questions, and the test designers randomly draw one question from each group. In the individual chapters of this study guide, where we look at the questions, you will see that the start of each question has a code that looks like **E1A01**. We decode this marking as

E1 — Element 4 (Extra), Subelement 1
A — Group A of Subelement 1
01 — Question 1 from Group A

This study guide uses the same format as the published question pool. On your exam, the wording of the questions and the answer choices will be the same, but the test producer will randomize both the order of the question groups and the answer choice order for each question. Before selecting the correct answer, carefully read each question and the four possible answers. As you will see, some of the answer options have similar wording.

Before the Exam Day

A Volunteer Examiner (VE) team administers license exams at prescheduled, publicly announced test sessions. Often, you will need to preregister with the team before taking the exam. Be sure to understand the date, time, location, and registration requirements, including any fees, for the session. Ask if this will be an "in-person" or "video-supervised" exam. If it is the latter, ask what equipment you need to take the exam.

If you require accommodations for any special needs, be sure to self-identify to the VE team the type of need and the necessary accommodations you require when you register for the exam. The VE team should permit you to have an accompanying adult or support animal at the session if needed. When you register, you should work with the VE team on the availability of sound-blocking devices or special room requirements, if needed.

If you do not have a FCC Registration Number (FRN), you must get one online from the FCC at https://apps.fcc.gov/cores/userLogin.do before the exam. *Note:* you must have an e-mail address to register with the FCC.

Table 1: 2020 – 2024 Extra Question Pool Organization

Subelement	Content	Groups	Questions
E1	Commission's rules: operating standards, station restrictions, definitions, space and Earth stations, Volunteer Examiner program, miscellaneous rules	6	75
E2	Operating procedures: space operations, telvision practices, opeerating methods	5	61
E3	Radio wave propagation: special methods, ionospheric effects, atmospheric and space weather effects	3	40
E4	Amateur practices: test equipment, measurement techniques, receiver characteristics, noise suppression and interference	5	60
E5	Electrical principles: resonance and Q, time constants and phase, phasors, AC and RF energy in circuits	4	55
E6	Circuit components: semiconductor devices, diodes, digital ICs, inductors, analog ICs, electro-optics	6	70
E7	Practical circuits: digital circuits, amplifiers, filters and matching networks, power supplies, DSP filtering, active filters, oscillators	8	108
E8	Signals and emissions: AC waveforms, modulation and demodulation, digital signals and effects	4	45
E9	Antennas and transmission lines: antenna parameters, antenna patterns, wire antennas, Yagi antennas, matching, transmission lines, Smith charts, receiving antennas	8	96
E0	Safety: RF hazards	1	11

Exam Day

The exam will only cover the Element 4 questions from the Extra Class license pool. You have no more amateur radio exams you can take in the US.

You will need to complete the NCVEC Quick-Form 605 at the session; see `http://www.arrl.org/files/file/VEs/NCVEC_Form_605_July_2022_INTER ACTIVE.pdf` and `http://www.arrl.org/fcc-qualification-question`. To complete this form, you will need to bring
- a copy of your current license
- your valid e-mail address
- your FRN

What else do you need to bring for the exam? Be sure to have the following items physically with you when the exam starts:
- several sharpened, number-two pencils, an eraser, and a pen
- the VE examination fee to cover the expenses of the exam (ask the VE team if cash, check, or money order is preferred); Note: the VEs receive no pay from this fee or any other fees
- a photo ID or other valid forms of identification (ask the examination team what alternatives they accept if a photo ID, such as a driver's license, is not available)
- you may also wish to bring a calculator whose memory you can prove you have cleared (the VE team will check for this); be sure that there are no exam-related formulas or data stored in the calculator's memory or programming
- you generally will not be allowed to use cell phones, smartphones, tablet computers, or similar electronic devices during the session

Verify this list with the examination team when you register or before the testing session begins to ensure the session has no other restrictions. See `http://www.arrl.org/what-to-bring-to-an-exam-session` for more information about the exam session.

You will have all the time necessary to complete the exam. Do not rush. Read each question carefully, and be sure to indicate the correct answer. There is no penalty for guessing. If you must guess the answer, try to eliminate as many choices as possible for the question. Then select the option that seems the most correct to you.

Good luck with the exam. I hope to meet you on the bands one day!

Chapter 1

E1 — COMMISSION RULES

1.1   Introduction

As we did in the Technician and General Class study guides, we begin our Extra Class study with the rules and regulations that govern the Amateur Radio Service. Because the Extra Class license holder has broader operating frequency privileges with worldwide access, there will be more detailed questions on permitted operations. The questions will frequently refer to the Federal Communications Commission (FCC) rules from Part 97 of the Code of Federal Regulations (CFR). You can find the current text for Part 97 at `https://www.ec` `fr.gov/current/title-47/chapter-I/subchapter-D/part-97?toc=1`.

As we look at Part 97 in more detail, you may need to make study cards to remember all the specific rules that the test can quiz you on. As we look at the questions and answers in this chapter, we will see quotes from Part 97 that are relevant. Next to the questions, the question pool designers also list the specific section in Part 97 where they drew the question. This Part 97 information will not appear on the exam questions – it is only for study purposes.

The *Commission Rules* subelement has the following question groups:
  A. Operating standards
  B. Station restrictions and special operations
  C. Definitions and restrictions
  D. Amateur space and Earth stations
  E. Volunteer examiner program
  F. Miscellaneous rules
Subelement 1 will generate six questions on the Extra examination.

1.2   Radio Engineering Concepts

Frequency Bands   The Extra Class frequency band access is easier to remember than the General and the Technician Class band access because Extra Class

Table 1.1: Non-CW Extra Class Emission Permissions in Region 2.

Band (m)	Permissions	Emissions
2200	135.7–137.8 kHz	RTTY, data, phone, image
630	472–479 kHz	RTTY, data, phone, image
160	1800–2000 kHz	RTTY, data, phone, image
80	3.500–3.600 MHz	RTTY, data
75	3.600–4.000 MHz	phone, image
60	5.3305–5.4064 MHz	phone, RTTY, data
40	7.000–7.125 MHz	RTTY, data
	7.125–7.300 MHz	phone, image
30	10.100–10.150 MHz	RTTY, data
20	14.000–14.150 MHz	RTTY, data
	14.150–14.350 MHz	phone, image
17	18.068–18.110 MHz	RTTY, data
	18.110–18.168 MHz	phone, image
15	21.000–21.200 MHz	RTTY, data
	21.200–21.450 MHz	phone, image
12	24.890–24.930 MHz	RTTY, data
	24.930–24.990 MHz	phone, image
10	28.000–28.300 MHz	RTTY, data
	28.300–29.700 MHz	phone, image

operators have full access to each band. Table 1.1 summarizes the Extra Class privileges on the Low Frequency (LF), Medium Frequency (MF), and High Frequency (HF) bands that we need to know. See Part 97 for the full list of privileges on all bands. Note: the radio regulations allow Continuous Wave (CW) on all bands, so the table does not explicitly list that mode.

US Boundaries   As you well know, the Radio Frequency (RF) transmissions do not always stay within national boundaries. Additionally, usage can differ in adjacent countries. The United States (US) has two special border regions of concern: the one with Canada and the one with Mexico. Figure 1.1 shows the "Line A" regulatory region around the contiguous US states and the southern Canadian border. There is a similar "Line C" for the Alaska/Canada border. Figure 1.2 illustrates the coordination region around the US/Mexico border. The Extra Class examination has a question about Line A.

Figure 1.1: Line A along the contiguous US states and the Canadian southern border. Courtesy FCC.

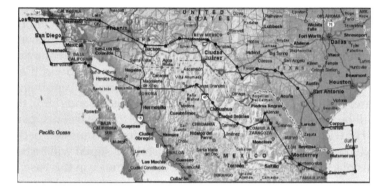

Figure 1.2: The coordination region along the US/Mexican border. Courtesy NTIA.

1.3 E1A - Operating Standards

1.3.1 Overview

The *Operating Standards* question group in Subelement E1 tests you on the allowed Extra Class MF and HF operating bands and the emissions permitted on those bands. The *Operating Standards* group covers topics such as

- Frequency privileges
- Automatic message forwarding
- Stations aboard ships or aircraft
- Power restrictions on the 630 and 2200-meter bands

The test producer will select one of the 14 questions in this group for your exam.

1.3.2   Questions

E1A01 [97.305, 97.307(b)] Which of the following carrier frequencies is illegal for LSB AFSK emissions on the 17 meter band RTTY and data segment of 18.068 to 18.110 MHz?
  A. 18.068 MHz
  B. 18.100 MHz
  C. 18.107 MHz
  D. 18.110 MHz

If we examine Table 1.1, we see that the 18.068 to 18.110 MHz region of the 17-meter band can be used for Radio TeleType (RTTY) and data. The Audio Frequency Shift Keying (AFSK) modulation technique is a common way to transmit data messages. However, Part 97 stipulates that "[e]missions resulting from modulation must be confined to the band or segment available to the control operator. Emissions outside the necessary bandwidth must not cause splatter or keyclick interference to operations on adjacent frequencies." In this case, since the operator wishes to use Lower Side Band (LSB), the AFSK signal will be below the carrier frequency during transmission. Operating at 18.068 MHz will cause an unallowed out-of-band transmission because the carrier will be at the band edge, and the transmitted signal will be below the edge. This placement makes **Answer A** the correct choice in this case.

E1A02 [97.301, 97.305] When using a transceiver that displays the carrier frequency of phone signals, which of the following displayed frequencies represents the lowest frequency at which a properly adjusted LSB emission will be totally within the band?
  A. The exact lower band edge
  B. 300 Hz above the lower band edge
  C. 1 kHz above the lower band edge
  D. 3 kHz above the lower band edge

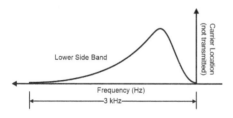

Figure 1.3: The LSB spectrum relative to the carrier frequency.

As we just saw, Part 97 states that any station's emissions "resulting from modulation must be confined to the band or segment available to the control operator." Therefore, if the transmission is near the band edge, the operator must know how close to the band edge the operator can tune the transmitter. Since LSB phone spectrum is approximately 3 kHz wide, as Figure 1.3 shows, the operator may tune the transmitter no lower than 3 kHz above the lower band edge, as in **Answer D**. The other choices allow the transmission to spill outside the band.

E1A03 [97.305, 97.307(b)] What is the maximum legal carrier frequency on the 20 meter band for transmitting USB AFSK digital signals having a 1 kHz bandwidth?
  A. 14.070 MHz
  B. 14.100 MHz
  C. 14.149 MHz
  D. 14.349 MHz

Using Table 1.1, the upper edge of the RTTY/data segment on the 20-m band is 14.150 MHz. This time, we are using Upper Side Band (USB) with the AFSK modulation having a bandwidth of 1 kHz. This means that the carrier must be at least 1 kHz below the band edge or 14.149 MHz as in **Answer C**.

E1A04 [97.301, 97.305] With your transceiver displaying the carrier frequency of phone signals, you hear a DX station calling CQ on 3.601 MHz LSB. Is it legal to return the call using lower sideband on the same frequency?
  A. Yes, because the DX station initiated the contact
  B. Yes, because the displayed frequency is within the 75 meter phone band segment
  C. No, the sideband will extend beyond the edge of the phone band segment
  D. No, U.S. stations are not permitted to use phone emissions below 3.610 MHz

Using Table 1.1, the lower edge of the phone segment on the 75-m is 3.600 MHz. This CQ is just above the permitted band limit in the US and less than 3 kHz from the band edge. If you were to answer the CQ, your return signal would extend beyond the permitted band edge, so you cannot legally answer the CQ as stated in **Answer C**.

E1A05 [97.313] What is the maximum power output permitted on the 60 meter band?
  A. 50 watts PEP effective radiated power relative to an isotropic radiator
  B. 50 watts PEP effective radiated power relative to a dipole
  C. 100 watts PEP effective radiated power relative to the gain of a half-wave dipole
  D. 100 watts PEP effective radiated power relative to an isotropic radiator

The rules in Part 97 state that "[n]o station may transmit with an effective radiated power (ERP) exceeding 100 W PEP on the 60 m band. For the purpose of computing ERP, the transmitter PEP will be multiplied by the antenna gain relative to a half-wave dipole antenna." **Answer C** captures this regulation. Be careful with Answer D because it has the proper power limit but the wrong reference antenna.

E1A06 [97.303(h)(1)] Where must the carrier frequency of a CW signal be set to comply with FCC rules for 60 meter operation?
- A. At the lowest frequency of the channel
- B. At the center frequency of the channel
- C. At the highest frequency of the channel
- D. On any frequency where the signal's sidebands are within the channel

Part 97 states that for CW emissions, "the carrier frequency is set to the center frequency," so **Answer B** is the right choice. All others are not in compliance with Part 97.

E1A07 [97.313(k)] What is the maximum power permitted on the 2200 meter band?
- A. 50 watts PEP
- B. 100 watts PEP
- C. 1 watt EIRP (Equivalent isotropic radiated power)
- D. 5 watts EIRP (Equivalent isotropic radiated power)

Part 97 states that "[n]o station may transmit in the 135.7-137.8 kHz (2200 m) band with a transmitter power exceeding 1.5 kW PEP or a radiated power exceeding 1 W EIRP." **Answer C** is the correct choice.

E1A08 [97.219] If a station in a message forwarding system inadvertently forwards a message that is in violation of FCC rules, who is primarily accountable for the rules violation?
- A. The control operator of the packet bulletin board station
- B. The control operator of the originating station
- C. The control operators of all the stations in the system
- D. The control operators of all the stations in the system not authenticating the source from which they accept communications

Operators trust that all message forwarding systems send messages with appropriate content and proper formats. The regulations in Part 97 state that for "stations participating in a message forwarding system, the control operator of the station originating a message is primarily accountable for any violation of the rules in this part contained in the message." **Answer B** conforms to the rules, so it is the correct choice for this question.

E1A09 [97.219] What action or actions should you take if your digital message forwarding station inadvertently forwards a communication that violates FCC rules?
   A. Discontinue forwarding the communication as soon as you become aware of it
   B. Notify the originating station that the communication does not comply with FCC rules
   C. Notify the nearest FCC Field Engineer's office
   D. All these choices are correct

Good amateur practice obliges the operator to fix the violation problem quickly. For this case, the Part 97 regulations state that the control operator is "responsible for discontinuing such communications once they become aware of their presence." In this situation, **Answer A** is the correct choice.

E1A10 [97.11] If an amateur station is installed aboard a ship or aircraft, what condition must be met before the station is operated?
   A. Its operation must be approved by the master of the ship or the pilot in command of the aircraft
   B. The amateur station operator must agree not to transmit when the main radio of the ship or aircraft is in use
   C. The amateur station must have a power supply that is completely independent of the main ship or aircraft power supply
   D. The amateur operator must have an FCC Marine or Aircraft endorsement on his or her amateur license

In Part 97, the regulations covering this situation state that "the installation and operation of an amateur station on a ship or aircraft must be approved by the master of the ship or pilot in command of the aircraft." **Answer A** covers this rule.

E1A11 [97.5] Which of the following describes authorization or licensing required when operating an amateur station aboard a U.S.-registered vessel in international waters?
   A. Any amateur license with an FCC Marine or Aircraft endorsement
   B. Any FCC-issued amateur license
   C. Only General Class or higher amateur licenses
   D. An unrestricted Radiotelephone Operator Permit

Part 97 states that the "station apparatus must be under the physical control of a person named in an amateur station license grant ... or a person authorized for alien reciprocal operation ... before the station may transmit on any amateur service frequency from any place that is ... Within 50 km of the Earth's surface and aboard any vessel or craft that is documented or registered in the United States". **Answer B** matches the regulations.

E1A12 What special operating frequency restrictions are imposed on slow scan TV transmissions?
   A. None; they are allowed on all amateur frequencies
   B. They are restricted to 7.245 MHz, 14.245 MHz, 21.345 MHz, and 28.945 MHz
   C. They are restricted to phone band segments
   D. They are not permitted above 54 MHz MHz

Slow-Scan Television (SSTV) is a form of image transmission that uses Single Sideband (SSB) modulation for the transmission, and it uses about the same bandwidth as a phone transmission. If you review Table 1.1, you will see that where image transmissions are allowed, they are in the same band segments as phone emissions. This restriction makes **Answer C** the correct choice.

E1A13 [97.5] Who must be in physical control of the station apparatus of an amateur station aboard any vessel or craft that is documented or registered in the United States?
   A. Only a person with an FCC Marine Radio license grant
   B. Any person holding an FCC issued amateur license or who is authorized for alien reciprocal operation
   C. Only a person named in an amateur station license grant
   D. Any person named in an amateur station license grant or a person holding an unrestricted Radiotelephone Operator Permit

Repeating the Part 97 regulation: "station apparatus must be under the physical control of a person named in an amateur station license grant on the ULS consolidated license database or a person authorized for alien reciprocal operation". This regulation makes **Answer B** the right choice.

E1A14 [97.313(l)] Except in some parts of Alaska, what is the maximum power permitted on the 630 meter band?
   A. 50 watts PEP
   B. 100 watts PEP
   C. 1 watt EIRP
   D. 5 watts EIRP

Part 97 states that "[n]o station may transmit in the 472-479 kHz (630 m) band with a transmitter power exceeding 500 W PEP or a radiated power exceeding 5 W EIRP, except that in Alaska, stations located within 800 kilometers of the Russian Federation may not transmit with a radiated power exceeding 1 W EIRP." Since the question is not asking about operating in Alaska, the correct choice is the 5 W in **Answer D**. Note: many of these antennas are very inefficient, hence the wide difference between the 500 W Peak Envelope Power (PEP) and the 5 W Effective Isotropic Radiated Power (EIRP).

1.4 E1B - Station Restrictions and Special Operations

1.4.1 Overview

The *Station Restrictions and Special Operations* question group in Subelement E1 tests you on various rules related to operating on the bands. The *Station Restrictions and Special Operations* group covers topics such as
- Restrictions on station location
- General operating restrictions
- Spurious emissions
- Antenna structure restrictions
- RACES operations

The test producer will select one of the 12 questions in this group for your exam.

1.4.2 Questions

E1B01 [97.3] Which of the following constitutes a spurious emission?
- A. An amateur station transmission made at random without the proper call sign identification
- B. A signal transmitted to prevent its detection by any station other than the intended recipient
- C. Any transmitted signal that unintentionally interferes with another licensed radio station
- D. An emission outside the signal's necessary bandwidth that can be reduced or eliminated without affecting the information transmitted

Part 97 defines a spurious emission as an "emission, or frequencies outside the necessary bandwidth of a transmission, the level of which may be reduced without affecting the information being transmitted." **Answer D** is the choice matching the regulations. The other options are not part of good amateur practice either.

E1B02 [97.307(f)(2)] Which of the following is an acceptable bandwidth for Digital Radio Mondiale (DRM) based voice or SSTV digital transmissions made on the HF amateur bands?
- A. 3 kHz
- B. 10 kHz
- C. 15 kHz
- D. 20 kHz

Part 97 states that "[n]o non-phone emission shall exceed the bandwidth of a communications quality phone emission of the same modulation type." Since the Digital Radio Mondial (DRM) user wishes to send phone or SSTV, they need to use the same bandwidth as SSB phone, or 3 kHz, as in **Answer A**.

E1B03 [97.13] Within what distance must an amateur station protect an FCC monitoring facility from harmful interference?
- A. 1 mile
- B. 3 miles
- C. 10 miles
- D. 30 miles

The FCC requires that a "station within 1600 m (1 mile) of an FCC monitoring facility must protect that facility from harmful interference." **Answer A** is the correct choice in this situation.

E1B04 [97.13, 1.1305-1.1319] What must be done before placing an amateur station within an officially designated wilderness area or wildlife preserve, or an area listed in the National Register of Historical Places?
- A. A proposal must be submitted to the National Park Service
- B. A letter of intent must be filed with the Environmental Protection Agency
- C. An Environmental Assessment must be submitted to the FCC
- D. A form FSD-15 must be submitted to the Department of the Interior

In Part 97, the regulations state that restrictions might occur if the station is placed "on land of environmental importance or that is significant in American history, architecture or culture." The FCC requires that the operator submit an Environmental Assessment for operating facilities in one of these designated areas, as given in **Answer C**.

E1B05 [97.3] What is the National Radio Quiet Zone?
- A. An area in Puerto Rico surrounding the Arecibo Radio Telescope
- B. An area in New Mexico surrounding the White Sands Test Area
- C. An area surrounding the National Radio Astronomy Observatory
- D. An area in Florida surrounding Cape Canaveral

Part 97 defines the National Radio Quiet Zone as the "area in Maryland, Virginia and West Virginia Bounded by 39[deg] 15[min] N on the north, 78[deg] 30[min] W on the east, 37[deg] 30[min] N on the south and 80[deg] 30[min] W on the west." This zone is the area surrounding the National Radio Astronomy Observatory, as given in **Answer C**.

E1B06 [97.15] Which of the following additional rules apply if you are installing an amateur station antenna at a site at or near a public use airport?
  A. You may have to notify the Federal Aviation Administration and register it with the FCC as required by Part 17 of FCC rules
  B. You must submit engineering drawings to the FAA
  C. You must file an Environmental Impact Statement with the EPA before construction begins
  D. You must obtain a construction permit from the airport zoning authority

Part 97 states that around an airport, owners "of certain antenna structures more than 60.96 meters (200 feet) above ground level at the site or located near or at a public use airport must notify the Federal Aviation Administration and register with the Commission". **Answer A** is the correct choice.

E1B07 [97.15] To what type of regulations does PRB-1 apply?
  A. Homeowners associations
  B. FAA tower height limits
  C. State and local zoning
  D. Use of wireless devices in vehicles

The FCC undertook PRB-1 to "delineate the limitations of local zoning and other local and state regulatory authority over Federally-licensed radio facilities." **Answer C** matches this intent.

E1B08 [97.121] What limitations may the FCC place on an amateur station if its signal causes interference to domestic broadcast reception, assuming that the receivers involved are of good engineering design?
  A. The amateur station must cease operation
  B. The amateur station must cease operation on all frequencies below 30 MHz
  C. The amateur station must cease operation on all frequencies above 30 MHz
  D. The amateur station must avoid transmitting during certain hours on frequencies that cause the interference

Part 97 states that if "the operation of an amateur station causes general interference to the reception of transmissions from stations operating in the domestic broadcast service when receivers of good engineering design, including adequate selectivity characteristics, are used to receive such transmissions, and this fact is made known to the amateur station licensee, the amateur station shall not be operated during the hours from 8 p.m. to 10:30 p.m., local time, and on Sunday for the additional period from 10:30 a.m. until 1 p.m., local time, upon the frequency or frequencies used when the interference is created." This regulation says that the control operator may not use those amateur frequencies causing the interference at the specified times mentioned in **Answer D**.

E1B09 [97.407] Which amateur stations may be operated under RACES rules?
  A. Only those club stations licensed to Amateur Extra Class operators
  B. Any FCC-licensed amateur station except a Technician Class
  C. Any FCC-licensed amateur station certified by the responsible civil defense organization for the area served
  D. Any FCC-licensed amateur station participating in the Military Auxiliary Radio System (MARS)

The Part 97 regulations state that no "station may transmit in RACES unless it is an FCC-licensed primary, club, or military recreation station and it is certified by a civil defense organization as registered with that organization, or it is an FCC-licensed RACES station." **Answer C** matches the regulations.

E1B10 [97.407] What frequencies are authorized to an amateur station operating under RACES rules?
  A. All amateur service frequencies authorized to the control operator
  B. Specific segments in the amateur service MF, HF, VHF and UHF bands
  C. Specific local government channels
  D. Military Auxiliary Radio System (MARS) channels

Part 97 states that the "frequency bands and segments and emissions authorized to the control operator are available to stations transmitting communications in RACES on a shared basis with the amateur service." making **Answer A** correct.

E1B11 [97.15] What does PRB-1 require of regulations affecting amateur radio?
  A. No limitations may be placed on antenna size or placement
  B. Reasonable accommodations of amateur radio must be made
  C. Amateur radio operations must be permitted in any private residence
  D. Use of wireless devices in a vehicle is exempt from regulation

Part 97 states that for amateur radio communications, state and local governments "must reasonably accommodate such communications and must constitute the minimum practicable regulation to accomplish the state or local authority's legitimate purpose." **Answer B** best summarizes the PRB-1 ruling.

E1B12 [97.303(b)] What must the control operator of a repeater operating in the 70 cm band do if a radiolocation system experiences interference from that repeater?
  A. Cease operation or make changes to the repeater to mitigate the interference
  B. File an FAA NOTAM (Notice to Airmen) with the repeater system's ERP, call sign, and six-character grid locator
  C. Reduce the repeater antenna HAAT (Height Above Average Terrain)
  D. All these choices are correct

Part 97 requires that "[a]mateur stations transmitting in the 70 cm band ... must not cause harmful interference to, and must accept interference from, stations authorized by the United States Government in the radiolocation service." **Answer A** is the correct response for the operator to take.

1.5 E1C - Definitions and Restrictions

1.5.1 Overview

The *Definitions and Restrictions* question group in Subelement E1 tests you on the control and operations of amateur stations. The *Definitions and Restrictions* group covers topics such as
- Local, automatic, and remote-control operation
- Band-specific regulations
- Operating in, and communicating with, foreign countries
- Spurious emission standards
- HF modulation index limit
- Bandwidth definition

The test producer will select one of the 14 questions in this group for your exam.

1.5.2 Questions

E1C01 [97.303] What is the maximum bandwidth for a data emission on 60 meters?
A. 60 Hz
B. 170 Hz
C. 1.5 kHz
D. 2.8 kHz

For this band, Part 97 says "[a]mateur operators shall ensure that their emissions do not occupy more than 2.8 kHz". **Answer D** matches this restriction.

E1C02 [97.117] Which of the following types of communications may be transmitted to amateur stations in foreign countries?
A. Business-related messages for non-profit organizations
B. Messages intended for users of the maritime satellite service
C. Communications incidental to the purpose of the amateur service and remarks of a personal nature
D. All these choices are correct

Answer C captures the Part 97 regulation that states "[t]ransmissions to a different country, where permitted, shall be limited to communications incidental to the purposes of the amateur service and to remarks of a personal character."

E1C03 [97.109(d)] How do the control operator responsibilities of a station under automatic control differ from one under local control?
  A. Under local control there is no control operator
  B. Under automatic control the control operator is not required to be present at the control point
  C. Under automatic control there is no control operator
  D. Under local control a control operator is not required to be present at a control point

The Part 97 rules allow for automatic control, which is the "use of devices and procedures for control of a station when it is transmitting so that compliance with the FCC Rules is achieved without the control operator being present at a control point." **Answer B** matches the wording of the regulations.

E1C04 What is meant by IARP?
  A. An international amateur radio permit that allows U.S. amateurs to operate in certain countries of the Americas
  B. The internal amateur radio practices policy of the FCC
  C. An indication of increased antenna reflected power
  D. A forecast of intermittent aurora radio propagation

As listed in Part 97, the International Amateur Radio Permit (IARP) is a "document issued pursuant to the terms of the Inter-American Convention on an International Amateur Radio Permit by a country signatory to that Convention, other than the United States." **Answer A** matches this wording, making it the correct response.

E1C05 [97.221(c)(1), 97.115(c)] When may an automatically controlled station originate third party communications?
  A. Never
  B. Only when transmitting RTTY or data emissions
  C. When agreed upon by the sending or receiving station
  D. When approved by the National Telecommunication and Information Administration

Part 97 permits third-party communications with the restriction that no "station may transmit third party communications while being automatically controlled except a station transmitting a RTTY or data emission." This part of the regulations appears to make Answer B the correct choice. However, Part 97 also restricts operations with an automatically-controlled digital station to times when the "station is responding to interrogation by a station under local or remote control." This second part makes **Answer A** the correct choice.

E1C06 Which of the following is required in order to operate in accordance with CEPT rules in foreign countries where permitted?
  A. You must identify in the official language of the country in which you are operating
  B. The U.S. embassy must approve of your operation
  C. You must bring a copy of FCC Public Notice DA 16-1048
  D. You must append "/CEPT" to your call sign

This question covers an arcane bit of information for European Conference of Postal and Telecommunications Administrations (CEPT): you will need the FCC Public Notice DA 16-1048, as in **Answer C**, for this operating permission.

E1C07 [97.3(a)(8)] At what level below a signal's mean power level is its bandwidth determined according to FCC rules?
  A. 3 dB
  B. 6 dB
  C. 23 dB
  D. 26 dB

Part 97 defines the signal's bandwidth as the "width of a frequency band outside of which the mean power of the transmitted signal is attenuated at least 26 dB below the mean power of the transmitted signal within the band." **Answer D** matches this definition.

E1C08 [97.213] What is the maximum permissible duration of a remotely controlled station's transmissions if its control link malfunctions?
  A. 30 seconds
  B. 3 minutes
  C. 5 minutes
  D. 10 minutes

Answer B matches the Part 97 regulation, which states that, in this case, the station's control operator must make provisions that "are incorporated to limit transmission by the station to a period of no more than 3 minutes in the event of malfunction in the control link."

E1C09 [97.307] What is the highest modulation index permitted at the highest modulation frequency for angle modulation below 29.0 MHz?
  A. 0.5
  B. 1.0
  C. 2.0
  D. 3.0

In Part 97, we learn that no "angle-modulated emission may have a modulation index greater than 1 at the highest modulation frequency." **Answer B** matches this regulation.

E1C10 [97.307] What is the permitted mean power of any spurious emission relative to the mean power of the fundamental emission from a station transmitter or external RF amplifier installed after January 1, 2003 and transmitting on a frequency below 30 MHz?
   A. At least 43 dB below
   B. At least 53 dB below
   C. At least 63 dB below
   D. At least 73 dB below

For an emitter matching this exact condition, Part 97 requires that the emission "must be at least 43 dB below the mean power of the fundamental emission." **Answer A** is the correct choice.

E1C11 [97.5] Which of the following operating arrangements allows an FCC-licensed U.S. citizen to operate in many European countries, and alien amateurs from many European countries to operate in the U.S.?
   A. CEPT agreement
   B. IARP agreement
   C. ITU reciprocal license
   D. All of these choices are correct

The CEPT agreement allows reciprocal agreements between US and European amateurs, which makes **Answer A** the correct response.

E1C12 [97.305(c)] On what portion of the 630 meter band are phone emissions permitted?
   A. None
   B. Only the top 3 kHz
   C. Only the bottom 3 kHz
   D. The entire band

If we examine Table 1.1, we will see that Part 97 permits phone emissions across the entire 630-meter band. This permission makes **Answer D** the right choice.

E1C13 [97.303(g)] What notifications must be given before transmitting on the 630 meter or 2200 meter bands?
   A. A special endorsement must be requested from the FCC
   B. An environmental impact statement must be filed with the Department of the Interior
   C. Operators must inform the Utilities Technology Council (UTC) of their call sign and coordinates of the station
   D. Operators must inform the FAA of their intent to operate, giving their call sign and distance to the nearest runway

Part 97 requires that before operating "amateur operators shall notify the Utilities Telecom Council (UTC) of their intent to operate by submitting their call signs, intended band or bands of operation, and the coordinates of their antenna's fixed location. " **Answer C** matches this regulation. **Note**: the Utilities Telecom Council is now called the Utilities Technology Council.

E1C14 [97.303(g)] How long must an operator wait after filing a notification with the Utilities Technology Council (UTC) before operating on the 2200 meter or 630 meter band?
  A. Operators must not operate until approval is received
  B. Operators may operate after 30 days, providing they have not been told that their station is within 1 km of PLC systems using those frequencies
  C. Operators may not operate until a test signal has been transmitted in coordination with the local power company
  D. Operations may commence immediately, and may continue unless interference is reported by the UTC

Part 97 further states that amateurs "will be permitted to commence operations after the 30-day period unless UTC notifies the station that its fixed location is located within one kilometer of PLC systems operating in the same or overlapping frequencies." **Answer B** is the right choice here.

1.6   E1D - Amateur Space and Earth Stations

1.6.1   Overview

The *Amateur Space and Earth Stations* question group in Subelement E1 quizzes you on satellite operations. The *Amateur Space and Earth Stations* group covers topics such as
  • Amateur space and Earth stations
  • Telecommand and telemetry rules
  • Identification of balloon transmissions
  • One-way communications
The test producer will select one of the 12 questions in this group for your exam.

1.6.2   Questions

E1D01 [97.3] What is the definition of telemetry?
  A. One-way transmission of measurements at a distance from the measuring instrument
  B. Two-way transmissions in excess of 1000 feet
  C. Two-way transmissions of data
  D. One-way transmission that initiates, modifies, or terminates the functions of a device at a distance

Answer A has the definition found in Part 97: telemetry is a "one-way transmission of measurements at a distance from the measuring instrument." Be careful with Answer D because it is the definition of telecommand.

E1D02 [97.211(b)] Which of the following may transmit special codes intended to obscure the meaning of messages?
   A. Telecommand signals from a space telecommand station
   B. Data containing personal information
   C. Auxiliary relay links carrying repeater audio
   D. Binary control characters

Part 97 says that a "telecommand station may transmit special codes intended to obscure the meaning of telecommand messages to the station in space operation." **Answer A** matches this condition.

E1D03 [97.3(a)(45)] What is a space telecommand station?
   A. An amateur station located on the surface of the Earth for communication with other Earth stations by means of Earth satellites
   B. An amateur station that transmits communications to initiate, modify or terminate functions of a space station
   C. An amateur station located in a satellite or a balloon more than 50 kilometers above the surface of the Earth
   D. An amateur station that receives telemetry from a satellite or balloon more than 50 kilometers above the surface of the Earth

Part 97 defines a telecommand station as an "amateur station that transmits communications to initiate, modify or terminate functions of a space station." **Answer B** matches this definition.

E1D04 [97.119(a)] Which of the following is required in the identification transmissions from a balloon-borne telemetry station?
   A. Call sign
   B. The output power of the balloon transmitter
   C. The station's six-character Maidenhead grid locator
   D. All these choices are correct

While a balloon reaches a high altitude, regulators do not generally classify it as a space station so the Earth station rule from Part 97 that the station "must transmit its assigned call sign" still holds. This regulation makes **Answer A** the right choice.

E1D05 [97.213(d)] What must be posted at the station location of a station being operated by telecommand on or within 50 km of the Earth's surface?
A. A photocopy of the station license
B. A label with the name, address, and telephone number of the station licensee
C. A label with the name, address, and telephone number of the control operator
D. All these choices are correct

For telecommand stations, Part 97 requires that a "photocopy of the station license and a label with the name, address, and telephone number of the station licensee and at least one designated control operator is posted in a conspicuous place at the station location." Answers A, B, and C are each part of the regulation, so **Answer D** is the best choice.

E1D06 [97.215(c)] What is the maximum permitted transmitter output power when operating a model craft by telecommand?
A. 1 watt
B. 2 watts
C. 5 watts
D. 100 watts

For model craft situations, Part 97 states that the "transmitter power must not exceed 1 W." **Answer A** matches the regulation.

E1D07 [97.207] Which HF amateur bands have frequencies authorized for space stations?
A. Only the 40, 20, 17, 15, 12, and 10 meter bands
B. Only the 40, 20, 17, 15, and 10 meter bands
C. Only the 40, 30, 20, 15, 12, and 10 meter bands
D. All HF bands

The frequency bands for a space station are listed in Part 97. "The following frequency bands and segments are authorized to space stations:
(1) The 17 m, 15 m, 12 m, and 10 m bands, 6 mm, 4 mm, 2 mm and 1 mm bands; and
(2) The 7.0-7.1 MHz, 14.00-14.25 MHz, 144-146 MHz, 435-438 MHz, 1260-1270 MHz, and 2400-2450 MHz, 3.40-3.41 GHz, 5.83-5.85 GHz, 10.45-10.50 GHz, and 24.00-24.05 GHz segments."
From this list, we can spot **Answer A** as the correct choice among those given.

E1D08 [97.207] Which VHF amateur bands have frequencies authorized for space stations?
A. 6 meters and 2 meters
B. 6 meters, 2 meters, and 1.25 meters
C. 2 meters and 1.25 meters
D. 2 meters

Using the previous question as a guide, we can see that only the 2-m band is in the permitted list, so **Answer D** is the correct choice among those given.

E1D09 [97.207] Which UHF amateur bands have frequencies authorized for space stations?
A. 70 cm only
B. 70 cm and 13 cm
C. 70 cm and 33 cm
D. 33 cm and 13 cm

Using the list from the Part 97 regulations, we can see that the 70 cm and 13-cm bands of **Answer B** match the list from the choices given.

E1D10 [97.211] Which amateur stations are eligible to be telecommand stations of space stations (subject to the privileges of the class of operator license held by the control operator of the station)?
A. Any amateur station designated by NASA
B. Any amateur station so designated by the space station licensee
C. Any amateur station so designated by the ITU
D. All of these choices are correct

Part 97 states that "[a]ny amateur station designated by the licensee of a space station is eligible to transmit as a telecommand station for that space station, subject to the privileges of the class of operator license held by the control operator." **Answer B** matches the rules.

E1D11 [97.209] Which amateur stations are eligible to operate as Earth stations?
A. Any amateur station whose licensee has filed a pre-space notification with the FCC's International Bureau
B. Only those of General, Advanced or Amateur Extra Class operators
C. Only those of Amateur Extra Class operators
D. Any amateur station, subject to the privileges of the class of operator license held by the control operator

This question is similar to the telecommand case. Part 97 specifies that "[a]ny amateur station may be an Earth station. A holder of any class operator license may be the control operator of an Earth station, subject to the privileges of the class of operator license held by the control operator." **Answer D** is the choice that matches the regulations.

E1D12 [97.207(e), 97.203(g)] Which of the following amateur stations may transmit one-way communications?
   A. A space station, beacon station, or telecommand station
   B. A local repeater or linked repeater station
   C. A message forwarding station or automatically controlled digital station
   D. All these choices are correct

Normally, amateur communications are two-way exchanges. However, Part 97 permits space stations, beacon stations, and telecommand stations to transmit one-way communications to achieve their purpose. **Answer A** matches Part 97.

1.7   E1E - Volunteer Examiner Program

1.7.1   Overview

The *Volunteer Examiner Program* question group in Subelement E1 tests you on aspects of the Volunteer Examiner program rules. The *Volunteer Examiner Program* group covers topics such as
   • Definitions, qualifications, accreditation,
   • Question pools, preparation and administration of exams
   • Documentation requirements
The test producer will select one of the 12 questions in this group for your exam.

1.7.2   Questions

E1E01 [97.527] For which types of out-of-pocket expenses do the Part 97 rules state that VEs and VECs may be reimbursed?
   A. Preparing, processing, administering and coordinating an examination for an amateur radio license
   B. Teaching an amateur operator license examination preparation course
   C. No expenses are authorized for reimbursement
   D. Providing amateur operator license examination preparation training materials

While Part 97 does not permit anyone to pay Volunteer Examiners (VEs) to administer examinations, Part 97 allows that "VEs and VECs may be reimbursed by examinees for out-of-pocket expenses incurred in preparing, processing, administering, or coordinating an examination for an amateur operator license." **Answer A** meets this regulation.

E1E02 [97.523] Who does Part 97 task with maintaining the pools of questions for all U.S. amateur license examinations?
  A. The VEs
  B. The FCC
  C. The VECs
  D. The ARRL

Part 97 states that the authorized "VECs must cooperate in maintaining one question pool for each written examination element." This regulation makes **Answer C** the correct choice for this question. You can find the current question pools for each license element in cyberspace at http://www.ncvec.org/.

E1E03 [97.521] What is a Volunteer Examiner Coordinator?
  A. A person who has volunteered to administer amateur operator license examinations
  B. A person who has volunteered to prepare amateur operator license examinations
  C. An organization that has entered into an agreement with the FCC to coordinate, prepare, and administer amateur operator license examinations
  D. The person who has entered into an agreement with the FCC to be the VE session manager

Part 97 specifies that "No organization may serve as a VEC unless it has entered into a written agreement with the FCC." **Answer C** matches this regulation.

E1E04 [97.509, 97.525] Which of the following best describes the Volunteer Examiner accreditation process?
  A. Each General, Advanced and Amateur Extra Class operator is automatically accredited as a VE when the license is granted
  B. The amateur operator applying must pass a VE examination administered by the FCC Enforcement Bureau
  C. The prospective VE obtains accreditation from the FCC
  D. The procedure by which a VEC confirms that the VE applicant meets FCC requirements to serve as an examiner

Part 97 states that the VE must be "accredited by the coordinating VEC" but does not lay out the details of the process. Therefore, the accreditation process is the procedure the Volunteer Examiner Coordinator (VEC) follows for the VE applicant, as in **Answer D**. The other choices are restrictions not found in Part 97.

E1E05 [97.503] What is the minimum passing score on all amateur operator license examinations?
    A. Minimum passing score of 70%
    B. Minimum passing score of 74%
    C. Minimum passing score of 80%
    D. Minimum passing score of 77%

Part 97 gives the minimum passing score for each element and it turns out to be 74 %, so **Answer B** is the correct choice.

E1E06 [97.509] Who is responsible for the proper conduct and necessary supervision during an amateur operator license examination session?
    A. The VEC coordinating the session
    B. The FCC
    C. Each administering VE
    D. The VE session manager

Part 97 regulations state that each "administering VE must be present and observing the examinee throughout the entire examination." This rule makes **Answer C** the correct choice.

E1E07 [97.509] What should a VE do if a candidate fails to comply with the examiner's instructions during an amateur operator license examination?
    A. Warn the candidate that continued failure to comply will result in termination of the examination
    B. Immediately terminate the candidate's examination
    C. Allow the candidate to complete the examination, but invalidate the results
    D. Immediately terminate everyone's examination and close the session

The Part 97 regulations state that the "administering VEs are responsible for the proper conduct and necessary supervision of each examination. The administering VEs must immediately terminate the examination upon failure of the examinee to comply with their instructions." **Answer B** matches the regulations.

E1E08 [97.509] To which of the following examinees may a VE not administer an examination?
    A. Employees of the VE
    B. Friends of the VE
    C. Relatives of the VE as listed in the FCC rules
    D. All of these choices are correct

The VEs must avoid the appearance of a conflict of interest. Therefore, Part 97 requires that no "VE may administer an examination to his or her spouse, children, grandchildren, stepchildren, parents, grandparents, stepparents, brothers, sisters, stepbrothers, stepsisters, aunts, uncles, nieces, nephews, and in-laws."

Answer C summarizes this regulation.

E1E09 [97.509] What may be the penalty for a VE who fraudulently administers or certifies an examination?
- A. Revocation of the VE's amateur station license grant and the suspension of the VE's amateur operator license grant
- B. A fine of up to $1000 per occurrence
- C. A sentence of up to one year in prison
- D. All of these choices are correct

The Part 97 regulation covering this violation states that this "may result in the revocation of the grant of the VE's amateur station license and the suspension of the grant of the VE's amateur operator license." **Answer A** matches this regulation.

E1E10 [97.509(h)] What must the administering VEs do after the administration of a successful examination for an amateur operator license?
- A. They must collect and send the documents to the NCVEC for grading
- B. They must collect and submit the documents to the coordinating VEC for grading
- C. They must submit the application document to the coordinating VEC according to the coordinating VEC instructions
- D. They must collect and send the documents to the FCC according to instructions

Part 97 indicates that "[a]fter the administration of a successful examination for an amateur operator license, the administering VEs must submit the application document to the coordinating VEC according to the coordinating VEC's instructions." **Answer C** is the right choice.

E1E11 [97.509(m)] What must the VE team do if an examinee scores a passing grade on all examination elements needed for an upgrade or new license?
- A. Photocopy all examination documents and forward them to the FCC for processing
- B. Three VEs must certify that the examinee is qualified for the license grant and that they have complied with the administering VE requirements
- C. Issue the examinee the new or upgrade license
- D. All these choices are correct

When an examinee receives a passing grade, Part 97 states that the "3 VEs must certify that the examinee is qualified for the license grant and that the VEs have complied with these administering VE requirements." **Answer B** captures this regulation.

E1E12 [97.509(j)] What must the VE team do with the application form if the examinee does not pass the exam?
   A. Return the application document to the examinee
   B. Maintain the application form with the VEC's records
   C. Send the application form to the FCC and inform the FCC of the grade
   D. Destroy the application form

When an examinee does not pass, Part 97 states that the VE team "must return the application document to the examinee and inform the examinee of the grade." **Answer A** is the correct choice.

1.8   E1F - Miscellaneous Rules

1.8.1   Overview

The *Miscellaneous Rules* question group in Subelement E1 tests you on other rules for operating on the bands. The *Miscellaneous Rules* group covers topics such as
   • External RF power amplifiers
   • Prohibited communications
   • Spread spectrum
   • Auxiliary stations
   • Canadian amateurs operating in the U.S.
   • Special Temporary Authority
   • Control operator of an auxiliary station
The test producer will select one of the 11 questions in this group for your exam.

1.8.2   Questions

E1F01 [97.305] On what frequencies are spread spectrum transmissions permitted?
   A. Only on amateur frequencies above 50 MHz
   B. Only on amateur frequencies above 222 MHz
   C. Only on amateur frequencies above 420 MHz
   D. Only on amateur frequencies above 144 MHz

Spread Spectrum (SS) transmissions are special cases. Part 97 says that "no SS modulation emission may be transmitted on any frequency where SS is not specifically authorized." Part 97 shows spread spectrum as being allowed on Ultra High Frequency (UHF) frequencies and higher. These are above 222 MHz, so **Answer B** is the right choice.

E1F02 [97.107] What privileges are authorized in the U.S. to persons holding an amateur service license granted by the government of Canada?
- A. None, they must obtain a U.S. license
- B. All privileges of the Extra Class license
- C. The operating terms and conditions of the Canadian amateur service license, not to exceed U.S. Extra Class license privileges
- D. Full privileges, up to and including those of the Extra Class license, on the 80, 40, 20, 15, and 10 meter bands

Part 97 grants the Canadian license holder the same privileges as in Canada "applicable rules of this part, but not to exceed the control operator privileges of an FCC-granted Amateur Extra Class operator license." This regulation makes **Answer C** the right choice.

E1F03 [97.315] Under what circumstances may a dealer sell an external RF power amplifier capable of operation below 144 MHz if it has not been granted FCC certification?
- A. It was purchased in used condition from an amateur operator and is sold to another amateur operator for use at that operator's station
- B. The equipment dealer assembled it from a kit
- C. It was imported from a manufacturer in a country that does not require certification of RF power amplifiers
- D. It was imported from a manufacturer in another country and was certificated by that country's government

The Part 97 regulation states that the sale is permitted when the "amplifier is purchased in used condition by an equipment dealer from an amateur operator and the amplifier is further sold to another amateur operator for use at that operator's station." **Answer A** matches this condition.

E1F04 [97.3] Which of the following geographic descriptions approximately describes "Line A"?
- A. A line roughly parallel to and south of the border between the U.S. and Canada
- B. A line roughly parallel to and west of the U.S. Atlantic coastline
- C. A line roughly parallel to and north of the border between the U.S. and Mexico
- D. A line roughly parallel to and east of the U.S. Pacific coastline

Figure 1.1 illustrates Line A. As can be seen, it is a line parallel to and south of the US-Canadian border for the most part, so **Answer A** is the right choice. Answer C corresponds to Figure 1.2 so it is incorrect.

E1F05 [97.303] Amateur stations may not transmit in which of the following frequency segments if they are located in the contiguous 48 states and north of Line A?
   A. 440 MHz - 450 MHz
   B. 53 MHz - 54 MHz
   C. 222 MHz - 223 MHz
   D. 420 MHz - 430 MHz

Part 97 stipulates that no "amateur station shall transmit from north of Line A in the 420-430 MHz segment." This rule makes **Answer D** the correct choice. Be careful with Answer A because it is in the right overall band, just not the right portion of the band.

E1F06 [1.931] Under what circumstances might the FCC issue a Special Temporary Authority (STA) to an amateur station?
   A. To provide for experimental amateur communications
   B. To allow regular operation on Land Mobile channels
   C. To provide additional spectrum for personal use
   D. To provide temporary operation while awaiting normal licensing

The FCC may grant a Special Temporary Authority (STA) to "conduct tests to determine necessary data for the preparation of an application for regular authorization". **Answer A** is the only circumstance that would meet the requirements for a STA.

E1F07 [97.113] When may an amateur station send a message to a business?
   A. When the total money involved does not exceed $25
   B. When the control operator is employed by the FCC or another government agency
   C. When transmitting international third-party communications
   D. When neither the amateur nor his or her employer has a pecuniary interest in the communications

Part 97 bars communications "in which the station licensee or control operator has a pecuniary interest, including communications on behalf of an employer." This regulation makes **Answer D** the right choice among those given.

E1F08 [97.113(c)] Which of the following types of amateur station communications are prohibited?
   A. Communications transmitted for hire or material compensation, except as otherwise provided in the rules
   B. Communications that have a political content, except as allowed by the Fairness Doctrine
   C. Communications that have a religious content
   D. Communications in a language other than English

Based on the previous question, you should be able to spot that **Answer A** is the one that matches the regulations in Part 97.

E1F09 [97.311] Which of the following conditions apply when transmitting spread spectrum emissions?
   A. A station transmitting SS emission must not cause harmful interference to other stations employing other authorized emissions
   B. The transmitting station must be in an area regulated by the FCC or in a country that permits SS emissions
   C. The transmission must not be used to obscure the meaning of any communication
   D. All of these choices are correct

Each statement made in Answers A, B, and C is correct, so the best choice is **Answer D**.

E1F10 [97.201] Who may be the control operator of an auxiliary station?
   A. Any licensed amateur operator
   B. Only Technician, General, Advanced or Amateur Extra Class operators
   C. Only General, Advanced or Amateur Extra Class operators
   D. Only Amateur Extra Class operators

The regulations in Part 97 state that a "holder of a Technician, Technician Plus, General, Advanced or Amateur Extra Class operator license may be the control operator of an auxiliary station, subject to the privileges of the class of operator license held." **Answer B** is the one that is closest to the regulations.

E1F11 [97.317] Which of the following best describes one of the standards that must be met by an external RF power amplifier if it is to qualify for a grant of FCC certification?
   A. It must produce full legal output when driven by not more than 5 watts of mean RF input power
   B. It must be capable of external RF switching between its input and output networks
   C. It must exhibit a gain of 0 dB or less over its full output range
   D. It must satisfy the FCC's spurious emission standards when operated at the lesser of 1500 watts or its full output power

There are many restrictions on RF power amplifiers in Part 97. The restrictions start with "the amplifier must satisfy the spurious emission standards ... when the amplifier is ... Operated at its full output power ". **Answer D** captures this part of the regulations.

Chapter 2

E2 — OPERATING PROCEDURES

2.1 Introduction

The *Operating Procedures* subelement covers much more than basic Continuous Wave (CW) or phone operating. It revisits amateur satellite operations, television messaging, and other amateur operating modes across multiple interest areas such as contesting. These operations questions build on many of the topics seen earlier in the Technician and General Class examination questions. The *Operating Procedures* subelement has the following question groups:

- A. Amateur radio in space
- B. Television practices
- C. Contest and DX operating methods
- D. VHF and UHF digital operating methods
- E. HF digital operating methods

Subelement 2 will generate five questions on the Extra examination.

2.2 Radio Engineering Concepts

Satellite Orbits We saw an introduction to amateur satellites in the Technician Class study guide and some initial questions in Section 1.6. Here, we will go into more technical depth with the questions. First, we need to know some basic information about characterizing the satellite's orbit. Figure 2.1 illustrates an orbital ground track for the Hubble Space Telescope (HST). This track covers three consecutive orbits. While the actual orbit is a closed ellipse, it appears to be open on this type of plot because the Earth is also rotating underneath as the satellite orbits. There are several key points to notice:

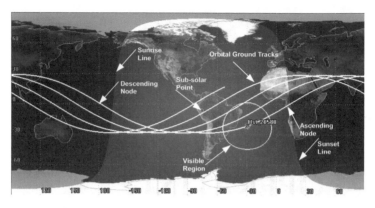

Figure 2.1: Orbital ground track for the Hubble Space Telescope (HST).

Ground Track A projection of the satellite's orbital path onto the Earth's surface

Node A point where the orbit crosses the Earth's equator

Ascending Node An orbital node when the satellite is traveling from south to north

Descending Node An orbital node when the satellite is traveling from north to south

Visible Region The region on the Earth's surface where the satellite is above the local horizon

There are several classes of satellite orbits that you need to be knowledgeable about:

LEO Low Earth Orbit — orbital altitudes below 2000 km (1,200 mi) with an orbital period of between 84 and 127 minutes

MEO Medium Earth Orbit — orbital altitudes above 2000 km (1,200 mi)) and below geostationary orbit

GEO Geosynchronous or Geostationary — an orbital altitude of 35 786 km (22,236 mi) and situated directly over the Equator with an orbital period of one day

HEO High Earth Orbit — orbital altitude above 35 786 km (22,236 mi) with an orbital period longer than one day

Satellite Operating Modes In the Extra Class exam questions, we will need to remember the upling/downlink arrangements for the satellite communications modes. Table 2.1 will help with these questions.

Radio Bands The Radio Regulations (RR) from the International Telecommunication Union (ITU) define Radio Frequency (RF) bands used in transmissions based on the frequency of the carrier. Table 2.2 lists the standard band designations. Table 2.3 shows how engineers in the communications industry also

Table 2.1: Amateur Satellite Communication Modes.

Mode	Satellite Receiving Band	Authorized Frequencies (MHz)	Satellite Transmitting Band	Authorized Frequencies (MHz)
V/H	VHF	144.30 – 144.50 145.80 – 146.00	HF	29.300 – 29.510
U/V	UHF	435.00 – 438.00	VHF	144.30 – 144.50 145.80 – 146.00
V/U	VHF	144.30 – 144.50 145.80 – 146.00	UHF	435.00 – 438.00
L/U	L-Band	1260 – 1270	UHF	435.00 – 438.00

Table 2.2: Standard Radio Band Designations.

Band Number	Designation	Frequency Range
3	Extremely Low Frequency (ELF)	<3 kHz
4	Very Low Frequency (VLF)	3 to 30 kHz
5	Low Frequency (LF)	30 to 300 kHz
6	Medium Frequency (MF)	300 to 3000 kHz
7	High Frequency (HF)	3 to 30 MHz
8	Very High Frequency (VHF)	30 to 300 MHz
9	Ultra High Frequency (UHF)	300 to 3000 MHz
10	Super High Frequency (SHF)	3 to 30 GHz
11	Extremely High Frequency (EHF)	30 to 300 GHz
12	Sub-millimeter	300 to 3000 GHz

use a letter designation for the bands. This naming convention comes from WWII radar designations, which were randomized for security reasons. Radio regulators do not recognize this letter designation for their purposes. The designations show some variability, depending upon the reference source cited.

2.3 E2A - Amateur Radio in Space

2.3.1 Overview

The *Amateur Radio in Space* question group in Subelement E2 tests you on the operational aspects of amateur satellite use. The *Amateur Radio in Space* group covers topics such as

Table 2.3: Engineering Frequency Band Designations.

Band	Frequency Range	Wavelength Range
VHF	30 to 300 MHz	10 to 1 m
UHF	300 to 1000 MHz	1 to 0.3 m
L	1 to 2 GHz	30 to 15 cm
S	2 to 4 GHz	15 to 7.5 cm
C	4 to 8 GHz	7.5 to 3.75 cm
X	8 to 12 GHz	3.75 to 2.5 cm
Ku	12 to 18 GHz	2.5 to 1.67 cm
K	18 to 27 GHz	1.67 to 1.1 cm
Ka	27 to 40 GHz	1.1 to 0.75 cm
V	40 to 75 GHz	0.75 to 0.4 cm
W	75 to 110 GHz	0.4 to 0.27 cm
mm	110 to 300 GHz	2.7 to 1.0 mm

Note: mm may be called G Band

- Amateur satellites
- Orbital mechanics
- Frequencies and modes
- Satellite hardware
- Satellite operations

The test producer will select one of the 13 questions in this group for your exam.

2.3.2   Questions

E2A01 What is the direction of an ascending pass for an amateur satellite?
  A. From west to east
  B. From east to west
  C. From south to north
  D. From north to south

Users reference the ascending and descending directions of a satellite's motion relative to the equator, as Figure 2.1 shows. Only the north or south motion is important and not the east-west motion. If the satellite is moving from south to north, then it makes an ascending pass, so **Answer C** is correct. Answer D is a descending pass.

E2A02 Which of the following occurs when a satellite is using an inverting linear transponder?
  A. Doppler shift is reduced because the uplink and downlink shifts are in opposite directions
  B. Signal position in the band is reversed
  C. Upper sideband on the uplink becomes lower sideband on the downlink, and vice versa
  D. All these choices are correct

Each statement in Answers A, B, and C is correct, thereby making **Answer D** the best choice for this question.

E2A03 How is the signal inverted by an inverting linear transponder?
  A. The signal is detected and remodulated on the reverse sideband
  B. The signal is passed through a non-linear filter
  C. The signal is reduced to I and Q components and the Q component is filtered out
  D. The signal is passed through a mixer and the difference rather than the sum is transmitted

If you remember about mixers from our General Class preparation, you know that the mixer produces the sum and the differences of the input frequencies. As **Answer D** says, the inverting transponder uses the difference rather than the sum from the mixer for the transmission. The other choices are to distract you.

E2A04 What is meant by the term "mode" as applied to an amateur radio satellite?
  A. Whether the satellite is in a low Earth or geostationary orbit
  B. The satellite's uplink and downlink frequency bands
  C. The satellite's orientation with respect to the Earth
  D. Whether the satellite is in a polar or equatorial orbit

This question is asking about the uplink and downlink communications modes in satellites. Table 2.1 summarizes the modes you need to remember. The first letter is the receiving band for the uplink and the second letter is the transmitting band for the downlink. These designations are from the satellite's point of view and not the Earth station using the satellite. **Answer B** captures the proper sense of the communications modes.

E2A05 What do the letters in a satellite's mode designator specify?
  A. Power limits for uplink and downlink transmissions
  B. The location of the ground control station
  C. The polarization of uplink and downlink signals
  D. The uplink and downlink frequency ranges

As we can see from Table 2.1, the modes designate the uplink and downlink frequency bands used in the communications. **Answer D** matches this usage.

E2A06 What are Keplerian elements?
  A. Parameters that define the orbit of a satellite
  B. Phase reversing elements in a Yagi antenna
  C. High-emission heater filaments used in magnetron tubes
  D. Encrypting codes used for spread spectrum modulation

Of the choices given, analysts only use the Keplerian elements of **Answer A** to define the satellite's orbit. For example, you can see the Keplerian elements for the HST satellite in Figure 2.2. The analyst usually enters the values into a computer program that performs the computations. The other choices are to distract you.

Figure 2.2: The Keplerian data for the HST orbital plot.

E2A07 Which of the following types of signals can be relayed through a linear transponder?
  A. FM and CW
  B. SSB and SSTV
  C. PSK and Packet
  D. All of these choices are correct

A linear transponder does not distort the carrier signal's amplitude, frequency, or phase. This property means that you can relay each of the signals mentioned in Answers A, B, and C through a linear transponder. **Answer D** the correct response.

E2A08 Why should effective radiated power to a satellite that uses a linear transponder be limited?
  A. To prevent creating errors in the satellite telemetry
  B. To avoid reducing the downlink power to all other users
  C. To prevent the satellite from emitting out-of-band signals
  D. To avoid interfering with terrestrial QSOs

Satellites have limited power, so they must manage their output transmission power effectively for all users. This restriction makes **Answer B** the best choice among those given.

E2A09 What do the terms "L band" and "S band" specify regarding satellite communications?
  A. The 23 centimeter and 13 centimeter bands
  B. The 2 meter and 70 centimeter bands
  C. FM and Digital Store-and-Forward systems
  D. Which sideband to use

If we refer to Table 2.3, we see that L Band covers the 23-cm amateur band, while S Band covers the 13-cm amateur band, so **Answer A** is the correct choice. Answer B is for the VHF and UHF bands. The other two options are not relevant in this context.

E2A10 What type of satellite appears to stay in one position in the sky?
  A. HEO
  B. Geostationary
  C. Geomagnetic
  D. LEO

This property is a quick way to describe a Geostationary (also called a Geosynchronous or GEO) satellite making **Answer B** is the right choice. Neither Low Earth Orbit (LEO) nor High Earth Orbit (HEO) satellites can appear stationary for an observer on the Earth. Answer C is to distract you.

E2A11 What type of antenna can be used to minimize the effects of spin modulation and Faraday rotation?
  A. A linearly polarized antenna
  B. A circularly polarized antenna
  C. An isotropic antenna
  D. A log-periodic dipole array

The linear, log periodic, and isotropic antenna types will not match the polarization shifts as the satellite passes from horizon to horizon. Hence, Answers A, C, and D are wrong and **Answer B** is correct. Circular polarization is not significantly affected by the satellite's spin or the Faraday rotation of the carrier.

E2A12 What is the purpose of digital store-and-forward functions on an amateur radio satellite?
  A. To upload operational software for the transponder
  B. To delay download of telemetry between satellites
  C. To store digital messages in the satellite for later download by other stations
  D. To relay messages between satellites

Because satellites pass over various users on the Earth as the satellite travels in its orbit, satellites can be convenient mailbox drops for messages. This operation is known as store-and-forward communications as described in **Answer C**, so this is the best choice for this question. As suggested in Answer D, the store-and-forward operation could eventually send the messages between satellites in some future protocol, but this does not happen regularly, so this is not a correct option.

E2A13 Which of the following techniques is normally used by low Earth orbiting digital satellites to relay messages around the world?
  A. Digipeating
  B. Store-and-forward
  C. Multi-satellite relaying
  D. Node hopping

LEO satellites cannot see large segments of the world at once (see the view region in Figure 2.1), so operators need to send and receive messages via a store-and-forward mode satellite "mailbox". This procedure makes **Answer B** the best choice among those given. Digipeating is an immediate retransmission mode, and if the receiver cannot see the satellite, the message is lost. Multi-satellite relaying does not exist in current amateur satellite systems. Node hopping works on the ground but not in amateur satellites, so this choice is also a distraction.

2.4 E2B - Television Practices

2.4.1 Overview

The *Television Practices* question group in Subelement E2 quizzes you on amateur television standards and operations. The *Television Practices* group covers fast-scan and slow-scan television standards and techniques. The test producer will select one of the 12 questions in this group for your exam.

2.4.2 Questions

E2B01 How many times per second is a new frame transmitted in a fast-scan (NTSC) television system?
  A. 30
  B. 60
  C. 90
  D. 120

This question is one whose answer you must memorize if you are unfamiliar with the National Television System Committee (NTSC) fast-scan TV standard (the type of TV used in analog commercial broadcasting before high-definition TV). The correct choice is 30 frames a second, as in **Answer A**. Be careful with

60 because that is the frequency of the Alternating Current (AC) in the standard power system. Answers C and D are to distract you.

E2B02 How many horizontal lines make up a fast-scan (NTSC) television frame?
  A. 30
  B. 60
  C. 525
  D. 1080

This question about fast-scan TV has an answer you also might need to memorize. The correct response is 525, as in **Answer C**. Be careful because 30 is for the previous question, and 60 is the power frequency. Answer D is associated with high-definition television.

E2B03 How is an interlaced scanning pattern generated in a fast-scan (NTSC) television system?
  A. By scanning two fields simultaneously
  B. By scanning each field from bottom to top
  C. By scanning lines from left to right in one field and right to left in the next
  D. By scanning odd numbered lines in one field and even numbered lines in the next

Here is another question on fast-scan TV. The TV system uses an alternating line pattern from one scan to the next, so **Answer D** is the correct choice. The other options do not describe the TV scanning process, so they are here to distract you.

E2B04 How is color information sent in analog SSTV?
  A. Color lines are sent sequentially
  B. Color information is sent on a 2.8 kHz subcarrier
  C. Color is sent in a color burst at the end of each line
  D. Color is amplitude modulated on the frequency modulated intensity signal

Answer A states the correct method for Slow-Scan Television (SSTV) of sending the lines sequentially. The other choices are technically incorrect statements.

E2B05 Which of the following describes the use of vestigial sideband in analog fast-scan TV transmissions?
  A. The vestigial sideband carries the audio information
  B. The vestigial sideband contains chroma information
  C. Vestigial sideband reduces bandwidth while allowing for simple video detector circuitry
  D. Vestigial sideband provides high frequency emphasis to sharpen the picture

A vestige implies a little bit of the original signal remains. In this case, Vestigial Side Band (VSB) is Single Sideband (SSB) with a piece of the undesired sideband left over. This technology achieves much of the bandwidth reduction goal of SSB over full Dual Sideband (DSB), but the technology required to demodulate the signal is much simpler than that needed for pure SSB. **Answer C** captures these technical goals.

E2B06 What is vestigial sideband modulation?
A. Amplitude modulation in which one complete sideband and a portion of the other are transmitted
B. A type of modulation in which one sideband is inverted
C. Narrow-band FM modulation achieved by filtering one sideband from the audio before frequency modulating the carrier
D. Spread spectrum modulation achieved by applying FM modulation following single sideband amplitude modulation

As we described in the previous question, VSB is SSB with a vestige of the second sideband also transmitted, as described in **Answer A**.

E2B07 What is the name of the signal component that carries color information in NTSC video?
A. Luminance
B. Chroma
C. Hue
D. Spectral Intensity

The chroma signal carries the color information, so look for **Answer B**. Luminance encodes the intensity information. Hue adjusts how the colors are displayed on the screen and is not a transmitted signal. Spectral intensity is not a TV signal parameter.

E2B08 What technique allows commercial analog TV receivers to be used for fast-scan TV operations on the 70 cm band?
A. Transmitting on channels shared with cable TV
B. Using converted satellite TV dishes
C. Transmitting on the abandoned TV channel 2
D. Using USB and demodulating the signal with a computer sound card

Be careful with Answer C because it does sound possible, but it is incorrect. The amateur fast-scan TV uses channels shared with cable TV as in **Answer A**. The others are to distract you.

E2B09 What hardware, other than a receiver with SSB capability and a suitable computer, is needed to decode SSTV using Digital Radio Mondiale (DRM)?
A. A special IF converter
B. A special front end limiter
C. A special notch filter to remove synchronization pulses
D. No other hardware is needed

Modern computer-based technology makes many modes easily accessible. As in **Answer D**, the user needs no additional equipment beyond that given in the question for Digital Radio Mondial (DRM)!

E2B10 What aspect of an analog slow-scan television signal encodes the brightness of the picture?
A. Tone frequency
B. Tone amplitude
C. Sync amplitude
D. Sync frequency

In SSTV, the tone frequency encodes the brightness of the picture, so **Answer A** is the correct choice.

E2B11 What is the function of the Vertical Interval Signaling (VIS) code sent as part of an SSTV transmission?
A. To lock the color burst oscillator in color SSTV images
B. To identify the SSTV mode being used
C. To provide vertical synchronization
D. To identify the call sign of the station transmitting

You need to remember that the Vertical Interval Signaling (VIS) code is to identify the user's SSTV mode, as indicated in **Answer B**.

E2B12 What signals SSTV receiving software to begin a new picture line?
A. Specific tone frequencies
B. Elapsed time
C. Specific tone amplitudes
D. A two-tone signal

As we saw earlier, SSTV uses tone frequencies to encode the picture for transmission. Therefore, the answer "specific tone frequencies," as listed in **Answer A**, is the method used to signal the beginning of a new picture line.

2.5   E2C - Contest and DX Operating Methods

2.5.1   Overview

The *Contest and DX Operating Methods* question group in Subelement E2 quizzes you on details and procedures of contesting and DX operating on the amateur bands. The *Contest and DX Operating Methods* group covers topics such as

* Contest and DX operating
* Remote operation techniques
* Cabrillo format
* QSLing
* RF network connected systems

The test producer will select one of the 12 questions in this group for your exam.

2.5.2   Questions

E2C01 What indicator is required to be used by U.S.-licensed operators when operating a station via remote control and the remote transmitter is located in the U.S.?

    A. / followed by the USPS two-letter abbreviation for the state in which the remote station is located

    B. /R# where # is the district of the remote station

    C. / followed by the ARRL Section of the remote station

    D. No additional indicator is required

Believe it or not, Part 97 does not require any unique indicator in this operating mode, so **Answer D** is the correct choice. The other options are to distract you.

E2C02 Which of the following best describes the term "self-spotting" in connection with HF contest operating?

    A. The often-prohibited practice of posting one's own call sign and frequency on a spotting network

    B. The acceptable practice of manually posting the call signs of stations on a spotting network

    C. A manual technique for rapidly zero beating or tuning to a station's frequency before calling that station

    D. An automatic method for rapidly zero beating or tuning to a station's frequency before calling that station

Most contest rules generally prohibit the participants from soliciting contacts. The practice of "self spotting" is a contact solicitation by posting one's information. The practice is frequently prohibited, as given in **Answer A**.

E2C03 From which of the following bands is amateur radio contesting generally excluded?
- A. 30 meters
- B. 6 meters
- C. 2 meters
- D. 33 centimeters

Generally, 30 m, with limited operating modes available and the need to avoid interference with other services, is not a contesting band. These restrictions make **Answer A** the right choice. The other bands are regularly involved in contesting throughout the year.

E2C04 Which of the following frequencies are sometimes used for amateur radio mesh networks?
- A. HF frequencies where digital communications are permitted
- B. Frequencies shared with various unlicensed wireless data services
- C. Cable TV channels 41 through 43
- D. The 60 meter band channel centered on 5373 kHz

Mesh networks often use re-purposed commercial wireless routers. The amateur network will share the unlicensed wireless frequencies with non-amateur users as in **Answer B**.

E2C05 What is the function of a DX QSL Manager?
- A. To allocate frequencies for DXpeditions
- B. To handle the receiving and sending of confirmation cards for a DX station
- C. To run a net to allow many stations to contact a rare DX station
- D. To relay calls to and from a DX station

If you start exchanging QSL cards with Distant (DX) stations, then you may need the services of one of the QSL bureaus. The QSL Manager oversees the handling of the QSL cards between amateur stations, so **Answer B** is the correct choice. The other activities do not fall to the QSL manager.

E2C06 During a VHF/UHF contest, in which band segment would you expect to find the highest level of SSB or CW activity?
- A. At the top of each band, usually in a segment reserved for contests
- B. In the middle of each band, usually on the national calling frequency
- C. In the weak signal segment of the band, with most of the activity near the calling frequency
- D. In the middle of the band, usually 25 kHz above the national calling frequency

Since there is no "reserved segment" for contesting, you should be able to spot Answer A as a distraction. Here a bit of reasoning will get you to the correct answer. For a contest, you will typically be trying to work weak signal stations, so going to the weak-signal part of the band, as mentioned in **Answer C**, is the right choice. The middle of the band, above the national calling frequency, might seem to be reasonable, but not for contests.

E2C07 What is the Cabrillo format?
  A. A standard for submission of electronic contest logs
  B. A method of exchanging information during a contest QSO
  C. The most common set of contest rules
  D. The rules of order for meetings between contest sponsors

Contest sponsors may request that participants submit contest logs in the Cabrillo standard data format. **Answer A** matches this use. Most modern contesting support software will generate this format for you as an option.

E2C08 Which of the following contacts may be confirmed through the U.S. QSL bureau system?
  A. Special event contacts between stations in the U.S.
  B. Contacts between a U.S. station and a non-U.S. station
  C. Repeater contacts between U.S. club members
  D. Contacts using tactical call signs

The QSL bureau handles QSL exchanges between United States (US) and DX stations. Each activity in Answers A, C, and D involves US-to-US exchanges, so these are not under the QSL bureau. Since **Answer B** deals with US/DX exchanges, it is the correct choice.

E2C09 What type of equipment is commonly used to implement an amateur radio mesh network?
  A. A 2 meter VHF transceiver with a 1200 baud modem
  B. An optical cable connection between the USB ports of 2 separate computers
  C. A standard wireless router running custom firmware
  D. A 440 MHz transceiver with a 9600 baud modem

Amateur radio mesh networks use modified Wi-Fi networking equipment, such as the standard wireless routers of **Answer C**. The other choices are distractions.

E2C10 Why might a DX station state that they are listening on another frequency?
  A. Because the DX station may be transmitting on a frequency that is prohibited to some responding stations
  B. To separate the calling stations from the DX station
  C. To improve operating efficiency by reducing interference
  D. All of these choices are correct

Each option in Answers A, B, and C is a reasonable cause for listening on another frequency, which makes **Answer D** the best choice.

E2C11 How should you generally identify your station when attempting to contact a DX station during a contest or in a pileup?
   A. Send your full call sign once or twice
   B. Send only the last two letters of your call sign until you make contact
   C. Send your full call sign and grid square
   D. Send the call sign of the DX station three times, the words this is, then your call sign three times

A "pileup" means that many operators are trying to work that DX station. To give everyone a fair shot at making the QSO, the operators must be efficient. In this case, only send your call sign once or twice, as in **Answer A**.

E2C12 What technique do individual nodes use to form a mesh network?
   A. Forward error correction and Viterbi codes
   B. Acting as store-and-forward digipeaters
   C. Discovery and link establishment protocols
   D. Custom code plugs for the local trunking systems

The mesh network requires the ability to discover users, nets, and link protocols, as in **Answer C**. The other choices are not needed for the mesh.

2.6   E2D - VHF/UHF Digital Operating Methods

2.6.1   Overview

The *VHF and UHF Digital Operating Methods* question group in Subelement E2 tests you on details of certain operating modes on these bands. The *VHF and UHF Digital Operating Methods* group covers topics such as
   • VHF and UHF digital modes and procedures
   • APRS
   • EME and meteor scatter procedures
The test producer will select one of the 11 questions in this group for your exam.

2.6.2   Questions

E2D01 Which of the following digital modes is designed for meteor scatter communications?
   A. WSPR
   B. MSK144
   C. Hellschreiber
   D. APRS

There are many digital modes, and some are designed especially for particular operating conditions. In this case, MSK144, as in **Answer B**, is such a digital mode intended for meteor scatter. WSPR is a weak-signal propagation reporter, Hellschreiber is a digital text exchange mode that builds up a print image one line at a time, and Automatic Packet Reporting System (APRS) is a general packet reporting system. The mode designers did not optimize these three for the short meteor scatter conditions, so they are not good choices. For more about MSK144, WSPR, and the other WSJT-X modes visit (https://wsjt.sourceforge.io/wsjtx-doc/wsjtx-main-2.6.1.html).

E2D02 Which of the following is a good technique for making meteor scatter contacts?
  A. 15-second timed transmission sequences with stations alternating based on location
  B. Use of special digital modes
  C. Short transmission with rapidly repeated call signs and signal reports
  D. All of these choices are correct

Meteor scatter communications require short messages with a quick handshake between stations. Operators can utilize each of the techniques given in Answers A, B, and C for this mode, so **Answer D** is the best choice for this question.

E2D03 Which of the following digital modes is especially useful for EME communications?
  A. MSK144
  B. PACTOR III
  C. Olivia
  D. JT65

Earth-Moon-Earth (EME) communications are low-strength signals when they arrive back at the Earth, so a low-signal mode, such as JT65, is an appropriate choice to support these communications. This property makes **Answer D** the best choice among those given. The other modes do not perform well in this weak-signal, high-noise transmission environment. MSK144 was designed for short QSOs using meteor scatter. PACTOR and Olivia were designed for traditional, Earth-bound QSOs.

E2D04 What technology is used to track, in real time, balloons carrying amateur radio transmitters?
  A. Ultrasonics
  B. Bandwidth compressed LORAN
  C. APRS
  D. Doppler shift of beacon signals

The APRS protocol (http://www.aprs.org/) allows for real-time position reporting based on inputs from a Global Positioning System (GPS) receiver. This feature makes **Answer C** the correct choice. Ultrasonics will help with medical imaging, but it is not a RF technique, so Answer A is not a good choice. Answers B and D are to distract you as well.

E2D05 What is one advantage of the JT65 mode?
A. Uses only a 65 Hz bandwidth
B. The ability to decode signals which have a very low signal-to-noise ratio
C. Easily copied by ear if necessary
D. Permits fast-scan TV transmissions over narrow bandwidth

The low signal-to-noise ratio reception ability is a key feature of JT65, so **Answer B** is the right choice. Each of the other answers is not a true characteristic of JT65.

E2D06 Which of the following describes a method of establishing EME contacts?
A. Time synchronous transmissions alternately from each station
B. Storing and forwarding digital messages
C. Judging optimum transmission times by monitoring beacons reflected from the Moon
D. High speed CW identification to avoid fading

Because of the time lags and the low signal strength in EME communications, operators synchronize their contacts, as in **Answer A**. The JT65 mode uses time synchronization in its protocol. The other suggestions are to distract you.

E2D07 What digital protocol is used by APRS?
A. PACTOR
B. 802.11
C. AX.25
D. AMTOR

The designer of the APRS system based it on the amateur AX.25 packet protocol, so **Answer C** is the right choice. Operators use the other protocols in amateur communications outside of APRS, so they are distractions here.

E2D08 What type of packet frame is used to transmit APRS beacon data?
A. Unnumbered Information
B. Disconnect
C. Acknowledgment
D. Connect

Here, you need to remember that beacon data are the information data the user sends periodically. The AX.25 protocol sends this type of data in an Unnumbered Information (UI) frame, so **Answer A** is the right choice. Beacons are not connected-mode services, so Answers B and D are wrong. Receivers do not acknowledge beacon data, so Answer C is not a good choice.

E2D09 What type of modulation is used for JT65 contacts?
A. Multi-tone AFSK
B. PSK
C. RTTY
D. IEEE 802.11

JT65 uses multi-tone Audio Frequency Shift Keying (AFSK), so **Answer A** is the right choice. Operators use the other choices in other amateur activities, but not JT65.

E2D10 How can an APRS station be used to help support a public service communications activity?
A. An APRS station with an emergency medical technician can automatically transmit medical data to the nearest hospital
B. APRS stations with General Personnel Scanners can automatically relay the participant numbers and time as they pass the check points
C. An APRS station with a Global Positioning System unit can automatically transmit information to show a mobile station's position during the event
D. All of these choices are correct

APRS transmitters send limited amounts of data, so they are not suitable for sending medical data in an emergency. APRS stations do not use General Personal Scanners (whatever they are). **Answer C** fits the APRS capabilities making it the best choice. Since Answers A and B are incorrect, Answer D is also wrong.

E2D11 Which of the following data are used by the APRS network to communicate your location?
A. Polar coordinates
B. Time and frequency
C. Radio direction finding spectrum analysis
D. Latitude and longitude

APRS uses standard longitude and latitude positional information from a GPS receiver making **Answer D** the correct choice. APRS does not use the other options.

2.7 E2E - HF Digital Operating Methods

2.7.1 Overview

The *HF Digital Operating Methods* question group in Subelement E2 tests you on your knowledge of operating with HF digital modes. The test producer will select one of the 13 questions in this group for your exam.

2.7.2 Questions

E2E01 Which of the following types of modulation is common for data emissions below 30 MHz?
   A. DTMF tones modulating an FM signal
   B. FSK
   C. Pulse modulation
   D. Spread spectrum

Amateurs frequently send data via Frequency Shift Keying (FSK) techniques on the HF bands so **Answer B** is correct. Operators more commonly encounter pulse modulation and Spread Spectrum (SS) on VHF and UHF bands and pulse modulation in optical communications. Dual Tone Multifrequency (DTMF) tones are for device control, and not data transmission.

E2E02 What do the letters FEC mean as they relate to digital operation?
   A. Forward Error Correction
   B. First Error Correction
   C. Fatal Error Correction
   D. Final Error Correction

The telecommunications industry has many forms of digital Forward Error Correction (FEC), so **Answer A** is the right choice. Be careful with the other options because you may select the wrong answer among the silly distractions.

E2E03 How is the timing of FT4 contacts organized?
   A. By exchanging ACK/NAK packets
   B. Stations take turns on alternate days
   C. Alternating transmissions at 7.5 second intervals
   D. It depends on the lunar phase

Alternating at 7.5-second intervals, as in **Answer C**, is the correct response for FT4. Answering on alternate days would not be efficient. Although other digital protocols use this technique, ACK/NAK packets are incorrect for FT4. The lunar phase choice is to make you smile.

E2E04 What is indicated when one of the ellipses in an FSK crossed-ellipse display suddenly disappears?
   A. Selective fading has occurred
   B. One of the signal filters is saturated
   C. The receiver has drifted 5 kHz from the desired receive frequency
   D. The mark and space signal have been inverted

Figure 2.3: Tuning ellipses for RTTY reception.

Many Radio TeleType (RTTY) software applications use a pair of ellipses to indicate tuning accuracy, as in Figure 2.3. With the propagation effect known as "frequency selective fading," a small region of the radio spectrum undergoes a deep fade as if the band applied a narrow filter. This fade is what can make one of the ellipses in the terminal unit disappear, so **Answer A** is the correct choice. The distraction answers are technobabble.

E2E05 Which of these digital modes does not support keyboard-to-keyboard operation?
   A. PACTOR
   B. RTTY
   C. PSK31
   D. MFSK

RTTY, PSK31, and Multiple Frequency Shifty Keying (MFSK) are all designed for keyboard-to-keyboard operations, so these are incorrect choices for this question. The PACTOR protocol of **Answer A** does not support this mode, so it is the correct choice here.

E2E06 What is the most common data rate used for HF packet?
   A. 48 baud
   B. 110 baud
   C. 300 baud
   D. 1200 baud

Since the question asks about the HF bands, the data rate will need to be low to match the available bandwidth. We saw different maximum rates for each band in the General Class studies. The 300-baud rate is "most common" because it is the maximum rate below 10 m, making **Answer C** the right choice. The 1200-baud rate is allowed on 10 m and above, so it is not "most common."

E2E07 Which of the following is a possible reason that attempts to initiate contact with a digital station on a clear frequency are unsuccessful?
A. Your transmit frequency is incorrect
B. The protocol version you are using is not supported by the digital station
C. Another station you are unable to hear is using the frequency
D. All these choices are correct

Each reason given in Answers A, B, and C is a possibility, so **Answer D** is the best choice to answer this question.

E2E08 Which of the following HF digital modes can be used to transfer binary files?
A. Hellschreiber
B. PACTOR
C. RTTY
D. AMTOR

Hellschreiber, RTTY, and Amateur Teleprinting Over Radio (AMTOR) were all designed to send printed text as their primary operating modes, so Answers A, C, and D are not correct choices here. Amateurs can use PACTOR for binary file transport, so **Answer B** is the best choice among those given.

E2E09 Which of the following HF digital modes uses variable-length coding for bandwidth efficiency?
A. RTTY
B. PACTOR
C. MT63
D. PSK31

In a sense, PSK31 works like Morse Code because the protocol assigns each character a length depending upon its frequency of use in text messages. This protocol results in variable-rate codes for characters, and **Answer D** is correct.

E2E10 Which of these digital modes has the narrowest bandwidth?
A. MFSK16
B. 170 Hz shift, 45 baud RTTY
C. PSK31
D. 300-baud packet

The MFSK16 transmission has a bandwidth of approximately 316 Hz. The 300-baud packet typically has a bandwidth of 500 Hz, and RTTY will be approximately 200 Hz. PSK31 has a bandwidth of 50 Hz, so it is the narrowest of those given, and **Answer C** is the right choice.

E2E11 What is the difference between direct FSK and audio FSK?
  A. Direct FSK applies the data signal to the transmitter VFO, while AFSK transmits tones via phone
  B. Direct FSK occupies less bandwidth
  C. Direct FSK can transmit faster baud rates
  D. Only direct FSK can be decoded by computer

Each statement in Answers B, C, and D is technically incorrect, so they are not good choices. Direct FSK directly modulates through the Variable Frequency Oscillator (VFO), while AFSK uses phone signals. **Answer A** is the right choice.

E2E12 How do ALE stations establish contact?
  A. ALE constantly scans a list of frequencies, activating the radio when the designated call sign is received
  B. ALE radios monitor an internet site for the frequency they are being paged on
  C. ALE radios send a constant tone code to establish a frequency for future use
  D. ALE radios activate when they hear their signal echoed by back scatter

The Automatic Link Enable (ALE) stations know the list of available operating frequencies, which they scan for participating stations, and then make contact when they find one. This property makes **Answer A** the correct choice. The others are incorrect procedures.

E2E13 Which of these digital modes has the fastest data throughput under clear communication conditions?
  A. AMTOR
  B. 170 Hz shift, 45 baud RTTY
  C. PSK31
  D. 300 baud packet

Under these conditions, the user needs the highest data rate for the fastest throughput (assumed to be error-free transmission). Of the choices given, the 300-baud packet has the highest throughput, so **Answer D** is the right choice. PSK31 has a data rate of 31 bits per second, so this is not a good choice. Both AMTOR and RTTY have data rates below 100 bps, so these are slower as well.

Chapter 3

E3 — RADIO WAVE PROPAGATION

3.1 Introduction

The *Radio Wave Propagation* subelement builds on the concepts discussed in the Technician and General Class studies. We will further investigate electromagnetic waves, propagation physics, and details on how the Radio Frequency (RF) radiation interacts with the various atmospheric layers. We will also see more on the nontraditional modes such as meteor scatter and Moon bounce communications. The *Radio Wave Propagation* subelement has the following question groups:

A. Electromagnetics
B. Radio wave propagation
C. Miscellaneous effects

Subelement 3 will generate three questions on the Extra examination.

3.2 Radio Engineering Concepts

Propagation Modes and Effects As we learned in our earlier studies, the Earth's atmosphere (see Figure 3.1), especially the ionosphere, has a significant influence on radio propagation. This subelement discusses several propagation modes and their effects. Here is a refresher on the many modes an amateur radio operator should know. Figure 3.2 illustrates many of these modes.

Direct Wave A propagation path directly connecting the transmitter and receiver antennas. The radio horizon limits the propagation on the High Frequency (HF) bands. This mode is most common for Very High Frequency (VHF) and higher bands.

Ground Wave At low frequencies, radio propagation along the Earth's surface

51

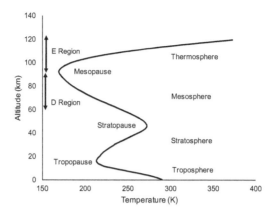

Figure 3.1: Earth's atmospheric layers including the D and E regions of the ionosphere.

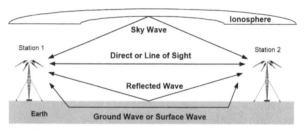

(a) Major modes of radio propagation through the atmosphere.

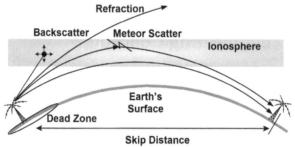

(b) Various sky wave propagation modes.

Figure 3.2: Ground, sky, and direct radio propagation modes.

is possible for a considerable distance. The distance falls off as frequency increases. Best for frequencies at Very Low Frequency (VLF) and below.

Sky Wave Radio propagation via reflection off the ionosphere. Operators also call this skip propagation. Best for frequencies below 50 MHz.

Meteor Scatter Radio waves reflect off the temporary meteor trails in the E-layer of the ionosphere. Best for VHF frequencies.

Auroral Backscatter Radio waves scatter in a backward direction from ionized regions associated with auroras. Best for HF and VHF bands.

Sporadic E Randomly occurring E-layer openings that tend to occur in mid-latitudes during daylight hours and in the summer months. Best for HF and VHF bands.

Tropospheric Ducting Radio waves achieve distant propagation by traveling in an atmospheric duct layer, usually caused by temperature inversion layers in the troposphere. Best for frequencies at VHF and above.

Tropospheric Scattering Also known as *troposcatter*, density turbulence in the atmosphere causes radio wave scattering. Best for frequencies at VHF and above.

Rain Scattering Radio waves are scattered by the rainfall. Best for frequencies around 10 GHz.

Airplane Scattering Radio wave reflection from the airplane's body. Best for frequencies at VHF and above.

Lightning Scattering A rare occurrence where the ionization path from the lightning can scatter radio waves. Sometimes observed at VHF and Ultra High Frequency (UHF) frequencies.

Diffraction Radio waves bend around relatively sharp obstructions to send the radiation into directions not possible by direct propagation. Diffraction characteristics depend on the radio wave's relative frequency and the distance to and size of the object.

Refraction The bending of an electromagnetic wave due to changes in the atmosphere's refraction index. The change is usually due to the changing density of the atmosphere along the propagation path.

Absorption Depending on the frequency, many different objects can absorb radio waves. Absorption can come from rain, atmospheric gases, foliage, buildings, etc.

3.3 E3A - Electromagnetics

3.3.1 Overview

The *Electromagnetics* question group in Subelement E3 tests you on the operational aspects of electromagnetic wave propagation. The *Electromagnetics* group covers topics such as

- Electromagnetic waves; circular polarization
- Earth-Moon-Earth and meteor scatter communications
- Microwave tropospheric and scatter propagation
- Aurora propagation
- Ionospheric propagation changes over the day

The test producer will select one of the 14 questions in this group for your exam.

3.3.2 Questions

E3A01 What is the approximate maximum separation measured along the surface of the Earth between two stations communicating by EME?
A. 500 miles, if the moon is at perigee
B. 2000 miles, if the moon is at apogee
C. 5000 miles, if the moon is at perigee
D. 12,000 miles, if the moon is visible by both stations

Given the great distance to the Moon, atmospheric refraction, mutual lunar visibility, etc., the best general answer is the 12,000 miles in **Answer D**. Answers A, B, and C are more typical for atmospheric propagation and not reflecting off the Moon.

E3A02 What characterizes libration fading of an EME signal?
A. A slow change in the pitch of the CW signal
B. A fluttery irregular fading
C. A gradual loss of signal as the sun rises
D. The returning echo is several hertz lower in frequency than the transmitted signal

Irregularities in the surface of the Moon and the changing viewing angle as the Moon moves relative to the radio operators cause libration fading. A fluttery fade in the Earth-Moon-Earth (EME) signal indicates libration fading is occurring, making **Answer B** the right choice. It does not change the pitch of the Continuous Wave (CW) signal, so Answer A is not correct. It is not a function of the sun angle, so Answer C is also incorrect. The libration motion does not cause a Doppler shift, so Answer D is not correct either.

E3A03 When scheduling EME contacts, which of these conditions will generally result in the least path loss?
A. When the moon is at perigee
B. When the moon is full
C. When the moon is at apogee
D. When the MUF is above 30 MHz

Path loss is proportional to the distance squared, so, naturally, the least path distance will cause the least path loss. The distance to the Moon does not correlate with the phase of the Moon, so Answer B is not relevant. The conditions in the ionosphere, which are not generally related to the Moon (even at Full Moon), determine the Maximum Usable Frequency (MUF), so Answer D is incorrect. When the Moon is at perigee, it is closest to the Earth, so **Answer A** is the right choice. When the Moon is at apogee, it is furthest away, and the path loss will be the largest.

E3A04 What do Hepburn maps predict?
  A. Sporadic E propagation
  B. Locations of auroral reflecting zones
  C. Likelihood of rain scatter along cold or warm fronts
  D. Probability of tropospheric propagation

Hepburn maps use tropospheric-condition forecasts to predict when favorable long-distance propagation will occur, especially via tropospheric ducting. **Answer D** matches this description. As in Answer C, the prediction of fronts is part of what goes into the maps, but the purpose is not in predicting rain events. The maps have nothing to do with auroras or sporadic E, so they are incorrect choices.

E3A05 Tropospheric propagation of microwave signals often occurs in association with what phenomenon?
  A. Grayline
  B. Lightning discharges
  C. Warm and cold fronts
  D. Sprites and jets

Tropospheric propagation enhancements correlate with warm and cold fronts as in **Answer C**. Grayline propagation correlates with the night/day line, which is incorrect. Lightning, sprites, and jets are associated with thunderstorm activity, not enhanced propagation, so these choices are wrong.

E3A06 What might help to restore contact when DX signals become too weak to copy across an entire HF band a few hours after sunset?
  A. Switch to a higher frequency HF band
  B. Switch to a lower frequency HF band
  C. Wait 90 minutes or so for the signal degradation to pass
  D. Wait 24 hours before attempting another communication on the band

The question indicates that the band conditions are deteriorating as is expected after sunset. One can continue to operate by changing frequencies. In this case, you should work a lower frequency because the MUF has moved lower as well, so **Answer B** is the correct choice. Answer A moves the frequency in the wrong direction, so this is not a good choice.

E3A07 Atmospheric ducts capable of propagating microwave signals often form over what geographic feature?
  A. Mountain ranges
  B. Forests
  C. Bodies of water
  D. Urban areas

Atmospheric ducting is often associated with inversion layers that can form over bodies of water, as in **Answer C**, so this is the right choice. The other options do not have this typical association, so they are incorrect.

E3A08 When a meteor strikes the Earth's atmosphere, a cylindrical region of free electrons is formed at what layer of the ionosphere?
  A. The E layer
  B. The F1 layer
  C. The F2 layer
  D. The D layer

Meteor trails form in the E layer, so **Answer A** is the right choice. The other options are to distract you.

E3A09 Which of the following frequency ranges is most suited for meteor scatter communications?
  A. 1.8 MHz - 1.9 MHz
  B. 10 MHz - 14 MHz
  C. 28 MHz - 148 MHz
  D. 220 MHz - 450 MHz

Operators make the majority of the meteor scatter contacts between the 10-m and 2-m bands, so **Answer C** is the best choice among those given. The other frequency bands are not optimal for meteor scatter contacts.

E3A10 Which type of atmospheric structure can create a path for microwave propagation?
  A. The jet stream
  B. Temperature inversion
  C. Wind shear
  D. Dust devil

Here we are back with atmospheric ducting. A duct also correlates with a temperature inversion layer, so **Answer B** is the correct choice. The other choices are to distract you.

E3A11 What is a typical range for tropospheric propagation of microwave signals?
- A. 10 miles to 50 miles
- B. 100 miles to 300 miles
- C. 1200 miles
- D. 2500 miles

You need to remember that tropospheric propagation for microwave signals typically extends over 100 to 300 miles making **Answer B** the correct choice.

E3A12 What is the cause of auroral activity?
- A. The interaction in the F2 layer between the solar wind and the Van Allen belt
- B. An extreme low-pressure area in the polar regions
- C. The interaction in the E layer of charged particles from the Sun with the Earth's magnetic field
- D. Meteor showers concentrated in the extreme northern and southern latitudes

Solar charged particles interacting with the Earth's magnetic field at the magnetic poles cause auroras. This interaction makes **Answer C** the right choice. The charged particles are part of the solar wind that interacts with the other ionospheric layers and the Van Allen belts. They do not cause auroras there, so Answer A is incorrect. Meteor showers only cause auroras in science fiction movies, so Answer D is also wrong. Answer B is to make you smile.

E3A13 Which of these emission modes is best for aurora propagation?
- A. CW
- B. SSB
- C. FM
- D. RTTY

Auroral propagation works best with a narrow-band signal, and CW has the narrowest bandwidth of the choices given, so **Answer A** is the best choice. The other options have wider bandwidths that the variable characteristics of auroral propagation can affect, so they are not good choices.

E3A14 What is meant by circularly polarized electromagnetic waves?
- A. Waves with an electric field bent into a circular shape
- B. Waves with a rotating electric field
- C. Waves that circle the Earth
- D. Waves produced by a loop antenna

The correct answer is that the circularly polarized waves have a rotating electric field, so **Answer B** is the right choice. All the other choices are silly distractions.

3.4 E3B - Radio Wave Propagation

3.4.1 Overview

The *Radio Wave Propagation* question group in Subelement E3 quizzes you on how the electromagnetic wave travels around the Earth. The *Radio Wave Propagation* group covers topics such as
- Transequatorial propagation
- Long path
- Ordinary and extraordinary waves
- Chordal hop
- Sporadic E mechanisms

The test producer will select one of the 11 questions in this group for your exam.

3.4.2 Questions

E3B01 What is transequatorial propagation?
- A. Propagation between two mid-latitude points at approximately the same distance north and south of the magnetic equator
- B. Propagation between points located on the magnetic equator
- C. Propagation between a point on the equator and its antipodal point
- D. Propagation between two stations at the same latitude

Transequatorial means across the Earth's magnetic equator, so Answers C and D are incorrect. **Answer A** gives the correct definition. Answer B is along the magnetic equator and not across it, so it is also incorrect.

E3B02 What is the approximate maximum range for signals using transequatorial propagation?
- A. 1000 miles
- B. 2500 miles
- C. 5000 miles
- D. 7500 miles

The correct choice here is 5000 miles as in **Answer C**. The others are to distract you.

E3B03 What is the best time of day for transequatorial propagation?
- A. Morning
- B. Noon
- C. Afternoon or early evening
- D. Late at night

Afternoons and early evenings are the best time for transequatorial propagation, so **Answer C** is the correct choice for this question. The other times are not optimal for this propagation mode, so they are not good choices.

E3B04 What is meant by the terms "extraordinary" and "ordinary" waves?
A. Extraordinary waves describe rare long-skip propagation compared to ordinary waves, which travel shorter distances
B. Independent waves created in the ionosphere that are elliptically polarized
C. Long-path and short-path waves
D. Refracted rays and reflected waves

When certain radio waves go through the ionosphere under the influence of the Earth's magnetic field, they can separate into two components known as the "ordinary" and "extraordinary" waves. **Answer B** reflects this description, so it is the right choice. The other options are distractions.

E3B05 Which amateur bands typically support long-path propagation?
A. Only 160 meters to 40 meters
B. Only 30 meters to 10 meters
C. 160 meters to 10 meters
D. 6 meters to 2 meters

Since VHF tends to have line-of-sight propagation, Answer D is not a good choice. Long-path propagation improves the Medium Frequency (MF) and HF bands. **Answer C** includes Answers A and B, so it is the best choice for this question.

E3B06 Which of the following amateur bands most frequently provides long-path propagation?
A. 80 meters
B. 20 meters
C. 10 meters
D. 6 meters

Answer D is incorrect because its frequency is too high, as we saw in the previous question. The 20-m band is a worldwide band open during daylight hours, so copying a signal along the "long path" is frequently possible, and **Answer B** is the best choice for this question.

E3B07 What happens to linearly polarized radio waves that split into ordinary and extraordinary waves in the ionosphere?
A. They are bent toward the magnetic poles
B. They become depolarized
C. They become elliptically polarized
D. They become phase locked

Splitting the RF wave into ordinary and extraordinary waves makes the result an elliptically-polarized wave, as in **Answer C**. The other choices are not technically correct.

E3B08 – question withdrawn

E3B09 At what time of year is sporadic E propagation most likely to occur?
  A. Around the solstices, especially the summer solstice
  B. Around the solstices, especially the winter solstice
  C. Around the equinoxes, especially the spring equinox
  D. Around the equinoxes, especially the fall equinox

Sporadic-E is most common around the solstices, especially the summer solstice, making **Answer A** the correct choice. Read the other options carefully because they are worded similarly but are incorrect.

E3B10 Why is chordal hop propagation desirable?
  A. The signal experiences less loss compared to multi-hop using Earth as a reflector
  B. The MUF for chordal hop propagation is much lower than for normal skip propagation
  C. Atmospheric noise is lower in the direction of chordal hop propagation
  D. Signals travel faster along ionospheric chords

With chordal hops, there is less attenuation than reflections that encounter the Earth's surface, which makes **Answer A** the correct choice.

E3B11 At what time of day is sporadic-E propagation most likely to occur?
  A. Only around sunset
  B. Only around sunset and sunrise
  C. Only in hours of darkness
  D. Any time

Sporadic-E propagation can occur at any time of the day, so **Answer D** is the right choice.

E3B12 What is the primary characteristic of chordal hop propagation?
  A. Propagation away from the great circle bearing between stations
  B. Successive ionospheric refractions without an intermediate reflection from the ground
  C. Propagation across the geomagnetic equator
  D. Signals reflected back toward the transmitting station

Chordal hop propagation is a RF signal that refracts around the Earth's ionosphere without encountering the ground, as in **Answer B**. Answer D is backscatter. Answer C is transequatorial propagation. Answer A is a distraction.

3.5 E3C - Miscellaneous Effects

3.5.1 Overview

The *Miscellaneous Effects* question group in Subelement E3 quizzes you on geomagnetic effects, space weather, and other propagation effects. The *Miscellaneous Effects* group covers topics such as
- Radio horizon
- Ground wave
- Propagation prediction techniques and modeling
- Effects of space weather parameters on propagation

The test producer will select one of the 15 questions in this group for your exam.

3.5.2 Questions

E3C01 What does the term "ray tracing" describe?
A. The process in which an electronic display presents a pattern
B. Modeling a radio wave's path through the ionosphere
C. Determining the radiation pattern from an array of antennas
D. Evaluating high voltage sources for x-rays

A ray trace is a model of how the RF signal propagates through the ionosphere, as in **Answer B**. The other choices are technically incorrect descriptions.

E3C02 What is indicated by a rising A or K index?
A. Increasing disruption of the geomagnetic field
B. Decreasing disruption of the geomagnetic field
C. Higher levels of solar UV radiation
D. An increase in the critical frequency

As we saw in our General Class studies, the rising A and K index means an increase in geomagnetic activity, so **Answer A** is correct. Be careful with Answer B because it is the opposite of the effect. Answers C and D are technically incorrect.

E3C03 Which of the following signal paths is most likely to experience high levels of absorption when the A index or K index is elevated?
A. Transequatorial
B. Polar
C. Sporadic E
D. NVIS

The particles from the Sun will be coming in at the Earth's magnetic poles. The polar paths will have the worst disruption, which makes **Answer B** the correct choice.

E3C04 What does the value of Bz (B sub z) represent?
A. Geomagnetic field stability
B. Critical frequency for vertical transmissions
C. Direction and strength of the interplanetary magnetic field
D. Duration of long-delayed echoes

Usually, B is a vector quantity that physicists use for the magnetic flux density. In this case, B_z is the direction and strength of the interplanetary magnetic field, so **Answer C** is the correct choice.

E3C05 What orientation of Bz (B sub z) increases the likelihood that incoming particles from the Sun will cause disturbed conditions?
A. Southward
B. Northward
C. Eastward
D. Westward

Here, B_z having a southward orientation indicates a higher likelihood of a disturbance making **Answer A** the right choice.

E3C06 By how much does the VHF/UHF radio horizon distance exceed the geometric horizon?
A. By approximately 15 percent of the distance
B. By approximately twice the distance
C. By approximately 50 percent of the distance
D. By approximately four times the distance

Atmospheric refraction makes the radio horizon appear to be 15 % farther than the geometric horizon, so the correct choice is **Answer A**. The others are to distract you.

E3C07 Which of the following descriptors indicates the greatest solar flare intensity?
A. Class A
B. Class B
C. Class M
D. Class X

A Class-X solar flare is the most intense, so **Answer D** is the right choice. Class M is a medium class, while Classes A and B are the weakest.

E3C08 What does the space weather term G5 mean?
A. An extreme geomagnetic storm
B. Very low solar activity
C. Moderate solar wind
D. Waning sunspot numbers

Scientists rate geomagnetic storms from G1 through G5, with G5 being the most intense. This definition makes **Answer A** the correct choice.

E3C09 How does the intensity of an X3 flare compare to that of an X2 flare?
A. 10 percent greater
B. 50 percent greater
C. Twice as great
D. Four times as great

The flare classes are factors of ten apart. Within the flare class, scientists divide the scale into finer divisions labeled from 1 to 9. An X2 flare is twice as powerful as an X1 flare, while an X3 flare is three times more powerful than an X1 flare. Since 3 is 50 % more than 2, **Answer B** is the correct choice. The others are math errors.

E3C10 What does the 304A solar parameter measure?
A. The ratio of x-ray flux to radio flux, correlated to sunspot number
B. UV emissions at 304 angstroms, correlated to solar flux index
C. The solar wind velocity at 304 degrees from the solar equator, correlated to solar activity
D. The solar emission at 304 GHz, correlated to x-ray flare levels

As in **Answer B**, the "304A" parameter measures Ultra Violet (UV) emissions at 304 angstroms, correlated to solar flux index. Be careful with the other answers having "304" in them because they are distractions.

E3C11 What does VOACAP software model?
A. AC voltage and impedance
B. VHF radio propagation
C. HF propagation
D. AC current and impedance

The Voice of America Coverage Analysis Program (VOCAP) is a standard software package for making HF propagation predictions as in **Answer C**. You can find more at https://www.voacap.com. The other choices are to distract you.

E3C12 How does the maximum distance of ground-wave propagation change when the signal frequency is increased?
- A. It stays the same
- B. It increases
- C. It decreases
- D. It peaks at roughly 14 MHz

Ground wave propagation is strongly frequency-dependent, and the propagation distance decreases with increasing frequency as in **Answer C**. Answer B is just the opposite of the correct effect. Answers A and D are technically wrong, so they are distractions.

E3C13 What type of polarization is best for ground-wave propagation?
- A. Vertical
- B. Horizontal
- C. Circular
- D. Elliptical

Typical ground wave propagation is vertically polarized, so the correct response is **Answer A**. The others are to see if you understand propagation basics.

E3C14 Why does the radio-path horizon distance exceed the geometric horizon?
- A. E-region skip
- B. D-region skip
- C. Due to the Doppler effect
- D. Downward bending due to density variations in the atmosphere

Radio waves undergo refraction, which is a bending caused by changes in the density of the atmosphere, as in **Answer D**. As we saw earlier, this makes the radio horizon about 15 % farther than the geometric horizon. Answers A and B are not relevant because they assume reflections off the ionosphere. The Doppler effect is for a frequency shift due to motion.

E3C15 What might be indicated by a sudden rise in radio background noise across a large portion of the HF spectrum?
- A. A temperature inversion has occurred
- B. A solar flare has occurred
- C. Increased transequatorial propagation likely
- D. Long-path propagation is occurring

Solar flares, as in **Answer B**, are often the cause of sudden increases in background noise, especially across multiple bands. The other options are not usually the result of noise, so they are incorrect.

Chapter 4

E4 — AMATEUR PRACTICES

4.1 Introduction

The *Amateur Practices* subelement brings us back to radio electronics circuitry and signal processing in much more depth than we saw on the Technician and General Class examination questions. This subelement covers signal sampling techniques, receiver noise, receiver performance measures, and details of test equipment and measurements. The *Amateur Practices* subelement has the following question groups:
  A. Test equipment
  B. Measurement techniques and limitations
  C. Receiver signal and noise performance characteristics
  D. Receiver hardware performance characteristics
  E. Noise suppression and interference
Subelement 4 will generate five questions on the Extra examination.

4.2 Radio Engineering Concepts

Oscilloscopes and Spectrum Analyzers Oscilloscopes and spectrum analyzers are complementary pieces of test equipment. The oscilloscope shows the signal in the time domain with the *x-axis* showing time and the *y-axis* showing amplitude. The spectrum analyzer shows the frequency domain representation of the signal by taking the Fourier transform of the time domain signal. The *x-axis* shows frequency and the *y-axis* shows amplitude for each frequency. Figure 4.1 illustrates both views.

Signal Sampling Digital equipment uses digital sampling electronics in devices such as digital oscilloscopes, spectrum analyzers, and computer sound cards. Figure 4.2(a) shows the general configuration for digital sampling. The process includes

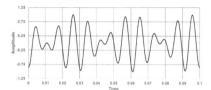

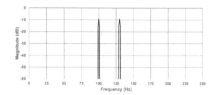

(a) Baseband time-domain signal as in an oscilloscope.

(b) Frequency domain signal as in a spectrum analyzer.

Figure 4.1: Time and frequency versions of the same signals as from an oscilloscope and a spectrum analyzer.

Sensor and Signal Conditioning Electronics A sensor responds to analog signal input and produces a proportional voltage or current on the output; the signal conditioning electronics produces an analog signal with proper characteristics for the next stages

Low Pass Fliter Limits the bandwidth of the input analog signal for proper sampling

Data Sampling Electronics Samples the analog signal at regular intervals and converts it to a proportional number at each sampling instant

The sampling electronics produce samples at a fixed rate. The Nyquist Theorem says that the sampling rate must be at least twice the frequency of the highest frequency component in the analog signal. Figure 4.2(b) shows the digital sampling electronics components. They are

Sample and Hold Amplifier Freezes the input signal for the duration of the analog-to-digital conversion process

Analog to Digital Converter Maps the input signal to a proportional digital code to represent the signal amplitude

What happens if the Analog-to-Digital Converter (ADC) does not sample quickly enough? If the ADC samples too slowly, signals above the Nyquist frequency map to lower frequencies, as Figure 4.3 shows. Engineers call this mapping aliasing, and it distorts the sampled signal. The anti-aliasing filter in Figure 4.2(a) cuts off frequencies above the Nyquist rate set by the sampling rate and prevents aliasing.

Two Port Networks Engineers often model electrical circuits based on their input and output characteristics measured at an input port and an output port rather than dealing with the details of the circuit components inside. Figure 4.4 shows this configuration. The inputs and outputs are related by

$$b_1 = S_{11}a_1 + S_{12}a_2$$
$$b_2 = S_{21}a_1 + S_{22}a_2$$

The S parameters are

S11 Input port voltage reflection coefficient

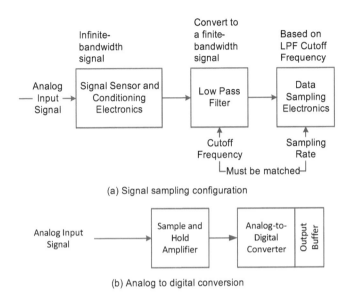

(a) Signal sampling configuration

(b) Analog to digital conversion

Figure 4.2: Digital sampling process in digital equipment and computer sound cards.

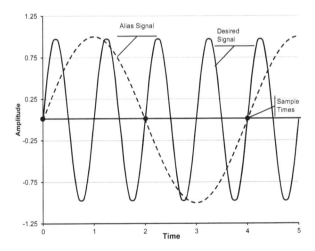

Figure 4.3: Signal aliasing from a higher frequency to a lower frequency.

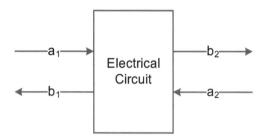

Figure 4.4: Two-port circuit model.

S12 Reverse voltage gain
S21 Forward voltage gain
S22 Output port voltage reflection coefficient

The S parameters can be used to express the circuit's gain and can also be used to estimate the Standing Wave Ratio (SWR).

Receiver Noise   When we try to estimate the performance of a receiver, we typically base it on the Signal-to-Noise Ratio (SNR). Engineers usually express the signal power as the average power. The noise power is the electronic noise in the receiver and the atmospheric noise the antenna picks up. Engineers express the receiver's noise as either a system temperature, T_{sys}, or a noise figure. Engineers compute the total noise power, N, by $N = kT_{sys}B$. Here, k is Boltzmann's constant ($1.380\,649 \times 10^{-23}$ W/(K − Hz) or -228.6 dBW/(K − Hz) in dB units) and B is the receiver's bandwidth. The system temperature is not the actual physical temperature of the receiver but is a parameter to characterize the noise. Often, engineers reference the noise to 290 K as a standard reference point.

The factor kT_{sys} is frequently referred to as the noise density, N_0, and has the units of W/Hz. The total noise is then $N = N_0B$. If one uses the standard reference temperature of 290 K, then the corresponding noise density is -174 dBm/Hz in dB units.

4.3   E4A - Test Equipment

4.3.1   Overview

The *Test Equipment* question group in Subelement E4 tests you on aspects of communications test equipment. The *Test Equipment* group covers topics such as

- Analog and digital instruments, spectrum analyzers, and oscilloscopes
- Antenna analyzers and RF measurements
- Computer-aided measurements

The test producer will select one of the 11 questions in this group for your exam.

4.3.2 Questions

E4A01 Which of the following limits the highest frequency signal that can be accurately displayed on a digital oscilloscope?
A. Sampling rate of the analog-to-digital converter
B. Amount of memory
C. Q of the circuit
D. All these choices are correct

The sampling rate is key to knowing the highest frequency the oscilloscope can sample correctly. The ADC converter inside the digital oscilloscope has a finite sampling rate, which sets the maximum frequency of the input signal. This rate makes **Answer A** the correct choice. The memory in the oscilloscope is important for determining the maximum sampling duration.

E4A02 Which of the following parameters would a spectrum analyzer display on the vertical and horizontal axes?
A. RF amplitude and time
B. RF amplitude and frequency
C. SWR and frequency
D. SWR and time

Figure 4.1 shows that the spectrum analyzer displays the signal's amplitude and frequency, as in **Answer B**. The oscilloscope shows the amplitude and time. The spectrum analyzer does not show SWR, so Answers C and D are incorrect.

E4A03 Which of the following test instruments is used to display spurious signals and/or intermodulation distortion products by an SSB transmitter?
A. A wattmeter
B. A spectrum analyzer
C. A logic analyzer
D. A time-domain reflectometer

Spurious signals are frequency-domain concepts, so a spectrum analyzer, as in **Answer B**, is the right choice. The wattmeter gives signal strength, the logic analyzer displays digital baseband signals, and a time-domain reflectometer identifies faults in cables, so these are not good choices.

E4A04 How is the compensation of an oscilloscope probe typically adjusted?
  A. A square wave is displayed and the probe is adjusted until the horizontal portions of the displayed wave are as nearly flat as possible
  B. A high frequency sine wave is displayed and the probe is adjusted for maximum amplitude
  C. A frequency standard is displayed and the probe is adjusted until the deflection time is accurate
  D. A DC voltage standard is displayed and the probe is adjusted until the displayed voltage is accurate

If you inject a square-wave signal, the horizontal portions will not stay flat if you have not properly adjusted the probe. The square wave has many harmonics, while a sinusoid has only one. This property makes the square wave a better choice to check the adjustment across a wide signal bandwidth, as indicated in **Answer A**.

E4A05 What is the purpose of the prescaler function on a frequency counter?
  A. It amplifies low-level signals for more accurate counting
  B. It multiplies a higher frequency signal so a low-frequency counter can display the operating frequency
  C. It prevents oscillation in a low-frequency counter circuit
  D. It divides a higher frequency signal so a low-frequency counter can display the input frequency

A prescaler is a frequency divider to allow a slower device to measure a high-frequency signal. This use makes **Answer D** the right answer for this question. Answer B is wrong because it does the opposite of the desired function. Answers A and C are electrically incorrect, so they are not good choices.

E4A06 What is the effect of aliasing on a digital oscilloscope caused by setting the time base too slow?
  A. A false, jittery low-frequency version of the signal is displayed
  B. All signals will have a DC offset
  C. Calibration of the vertical scale is no longer valid
  D. Excessive blanking occurs, which prevents display of the signal

As Figure 4.3 shows, the sampling hardware maps the high-frequency input signals to lower, incorrect frequencies when aliased. This distorted version of the signal makes **Answer A** correct. The other choices are to distract you.

E4A07 Which of the following is an advantage of using an antenna analyzer compared to an SWR bridge to measure antenna SWR?
   A. Antenna analyzers automatically tune your antenna for resonance
   B. Antenna analyzers do not need an external RF source
   C. Antenna analyzers display a time-varying representation of the modulation envelope
   D. All of these choices are correct

Each statement in Answers A and C is incorrect, so they are not good choices. That makes Answer D wrong as well. The antenna analyzer typically does not need an external RF source, so **Answer B** is the right choice.

E4A08 Which of the following measures SWR?
   A. A spectrum analyzer
   B. A Q meter
   C. An ohmmeter
   D. An antenna analyzer

As we saw in the last question, the antenna analyzer in **Answer D** is an excellent choice to measure the SWR of a beam antenna. The others are good pieces of test equipment, but not for this application.

E4A09 Which of the following is good practice when using an oscilloscope probe?
   A. Keep the signal ground connection of the probe as short as possible
   B. Never use a high-impedance probe to measure a low-impedance circuit
   C. Never use a DC-coupled probe to measure an AC circuit
   D. All of these choices are correct

Oscilloscope probes can pick up interfering signals that will influence measurements. A good way to minimize the effects of interference is to keep the ground lead as short as possible, as indicated in **Answer A**. Since Answers B and C are incorrect, Answer D is also incorrect.

E4A10 Which of the following displays multiple digital signal states simultaneously?
   A. Network analyzer
   B. Bit error rate tester
   C. Modulation monitor
   D. Logic analyzer

Of the options given, only the logic analyzer in **Answer D** can display multiple digital signals. The network analyzer is for analog circuit measurements. The bit error rate tester measures the performance of digital communications links, and the modulation monitor is for measuring carriers, so they are all incorrect choices.

E4A11 How should an antenna analyzer be connected when measuring antenna resonance and feed point impedance?
A. Loosely couple the analyzer near the antenna base
B. Connect the analyzer via a high-impedance transformer to the antenna
C. Loosely couple the antenna and a dummy load to the analyzer
D. Connect the antenna feed line directly to the analyzer's connector

Answer D shows the correct way to make the connection: directly connect the antenna feed line to the analyzer.

4.4 E4B - Measurement Techniques and Limitations

4.4.1 Overview

The *Measurement Techniques and Limitations* question group in Subelement E4 quizzes you on making measurements with test equipment and circuits. The *Measurement Techniques and Limitations* group covers topics such as
- Instrument accuracy and performance limitations
- Probes
- Techniques to minimize errors
- Measurement of Q
- Instrument calibration
- *S* parameters
- Vector network analyzers

The test producer will select one of the 11 questions in this group for your exam.

4.4.2 Questions

E4B01 Which of the following factors most affects the accuracy of a frequency counter?
A. Input attenuator accuracy
B. Time base accuracy
C. Decade divider accuracy
D. Temperature coefficient of the logic

If properly designed, the counter will only depend on the timing reference, so **Answer B** is the right choice for this question. Answers A, C, and D are incorrect for well-designed devices, so these are not good choices.

E4B02 What is the significance of voltmeter sensitivity expressed in ohms per volt?
  A. The full scale reading of the voltmeter multiplied by its ohms per volt rating will indicate the input impedance of the voltmeter
  B. When used as a galvanometer, the reading in volts multiplied by the ohms per volt rating will determine the power drawn by the device under test
  C. When used as an ohmmeter, the reading in ohms divided by the ohms per volt rating will determine the voltage applied to the circuit
  D. When used as an ammeter, the full scale reading in amps divided by ohms per volt rating will determine the size of shunt needed

The sensitivity units tell you how to understand the significance: multiply the sensitivity in ohms/volt by the input in volts. The result is in ohms, which is the measurement of impedance. This computation makes **Answer A** the correct choice.

E4B03 Which S parameter is equivalent to forward gain?
  A. S11
  B. S12
  C. S21
  D. S22

Figure 4.4 shows the S parameters. Forward gain corresponds to $S21$, so choose **Answer C**. $S11$ is the return loss term.

E4B04 Which S parameter represents input port return loss or reflection coefficient (equivalent to VSWR)?
  A. S11
  B. S12
  C. S21
  D. S22

Again, we refer to Figure 4.4. The return loss or SWR corresponds to $S11$, as in **Answer A**. Remember, $S21$ is the forward gain.

E4B05 What three test loads are used to calibrate an RF vector network analyzer?
  A. 50 ohms, 75 ohms, and 90 ohms
  B. Short circuit, open circuit, and 50 ohms
  C. Short circuit, open circuit, and resonant circuit
  D. 50 ohms through 1/8 wavelength, 1/4 wavelength, and 1/2 wavelength of coaxial cable

The correct test loads are short circuit, open circuit, and $50\,\Omega$, as in **Answer B**.

E4B06 How much power is being absorbed by the load when a directional power meter connected between a transmitter and a terminating load reads 100 watts forward power and 25 watts reflected power?
A. 100 watts
B. 125 watts
C. 25 watts
D. 75 watts

You should be able to compute this one in your head! If the transmitter's output is 100 W and the load reflects 25 W, then the load must absorb 75 W. This result matches **Answer D**. The other choices are to distract you.

E4B07 What do the subscripts of S parameters represent?
A. The port or ports at which measurements are made
B. The relative time between measurements
C. Relative quality of the data
D. Frequency order of the measurements

If we refer to Figure 4.4, the subscripts tell us at which port the user makes the measurement, so **Answer A** is the right choice. The other options are technically incorrect statements.

E4B08 Which of the following can be used to measure the Q of a series-tuned circuit?
A. The inductance to capacitance ratio
B. The frequency shift
C. The bandwidth of the circuit's frequency response
D. The resonant frequency of the circuit

As we will see in Chapter 5, the Quality Factor (Q) in a series resonant circuit is $Q = f_r/B$. Here, f_r is the resonant frequency of the circuit and B is the circuit's 3-dB bandwidth. While the resonant frequency of Answer D appears in the equation, it does not control the Q. The bandwidth controls the circuit's Q, making **Answer C** the best choice. The other choices are electrically incorrect.

E4B09 What is indicated if the current reading on an RF ammeter placed in series with the antenna feed line of a transmitter increases as the transmitter is tuned to resonance?
A. There is possibly a short to ground in the feed line
B. The transmitter is not properly neutralized
C. There is an impedance mismatch between the antenna and feed line
D. There is more power going into the antenna

Because the antenna and transmitter impedances match and we have maximum power transfer, this is the configuration we desire. In this case, the current to the antenna increases, and there is more power going to the antenna, as indicated in **Answer D**. The other choices are electrically incorrect statements.

E4B10 Which of the following methods measures intermodulation distortion in an SSB transmitter?
  A. Modulate the transmitter using two RF signals having non-harmonically related frequencies and observe the RF output with a spectrum analyzer
  B. Modulate the transmitter using two AF signals having non-harmonically related frequencies and observe the RF output with a spectrum analyzer
  C. Modulate the transmitter using two AF signals having harmonically related frequencies and observe the RF output with a peak reading wattmeter
  D. Modulate the transmitter using two RF signals having harmonically related frequencies and observe the RF output with a logic analyzer

Intermodulation distortion comes from nonlinear effects in circuits. One way to find these nonlinear effects is to excite the circuit with two Audio Frequency (AF) test tones that are not harmonically related to each other and have a spacing of approximately 1 kHz. It is essential that the tones not be harmonically related, as indicated in **Answer B**. If they are harmonically related, the measurement will not be accurate. Be careful with Answer A because it uses the wrong input signal type.

E4B11 Which of the following can be measured with a vector network analyzer?
  A. Input impedance
  B. Output impedance
  C. Reflection coefficient
  D. All these choices are correct

Answer D is the correct choice because we can measure each parameter given in Answers A, B, and C with a vector network analyzer.

4.5 E4C - Receiver Signal and Noise Performance

4.5.1 Overview

The *Receiver Signal and Noise Performance* question group in Subelement E4 tests you on the properties of receivers. The *Receiver Signal and Noise Performance* group covers topics such as
  • The receiver performance characteristics of phase noise, noise floor, image rejection, signal-to-noise-ratio, noise figure, and reciprocal mixing
  • Selectivity and Minimum Discernible Signal
  • Effects of SDR receiver non-linearity
  • The use of attenuators at low frequencies
The test producer will select one of the 15 questions in this group for your exam.

4.5.2   Questions

E4C01 What is an effect of excessive phase noise in a receiver's local oscillator?
A. It limits the receiver's ability to receive strong signals
B. It can affect the receiver's frequency calibration
C. It decreases receiver third-order intercept point
D. It can combine with strong signals on nearby frequencies to generate interference

Generally, the phase noise will not significantly affect the receiver's ability to receive strong signals. The receiver's frequency calibration is more determined by the receiver's overall design and not phase noise. Similarly, the third-order intercept point is more sensitive to the amplifiers than phase noise. Phase noise will affect the receiver's ability to demodulate signals, especially with weak signals near strong signals, making **Answer D** the correct choice.

E4C02 Which of the following receiver circuits can be effective in eliminating interference from strong out-of-band signals?
A. A front-end filter or pre-selector
B. A narrow IF filter
C. A notch filter
D. A properly adjusted product detector

The receiver's initial Radio Frequency (RF) circuitry is key to having a clean signal for demodulation. Therefore, we can mitigate interference by proper front-end filters or a preselector, as in **Answer A**. The other options will not effectively remove the interference because the receiver will be processing the signal before those options remove the interference, so they are not good choices.

E4C03 What is the term for the suppression in an FM receiver of one signal by another stronger signal on the same frequency?
A. Desensitization
B. Cross-modulation interference
C. Capture effect
D. Frequency discrimination

The Frequency Modulation (FM) capture effect is a good thing, provided that the receiver captures the signal you want to receive. The receiver locking onto the strongest signal is the correct definition of FM capture, as in **Answer C**. The definition of desensitization is the reduction of receiver sensitivity by a signal, so it is not the same effect. Cross-modulation interference is not unique to FM receivers, so Answer B is not the best choice. The frequency discriminator is the FM demodulator, so Answer D is not correct either.

E4C04 What is the noise figure of a receiver?
A. The ratio of atmospheric noise to phase noise
B. The ratio of the noise bandwidth in hertz to the theoretical bandwidth of a resistive network
C. The ratio of thermal noise to atmospheric noise
D. The ratio in dB of the noise generated by the receiver to the theoretical minimum noise

The noise figure is the ratio of the receiver's noise to a theoretical minimum when expressed in dB units. This definition makes **Answer D** the correct choice. The other options are to distract you.

E4C05 What does a receiver noise floor of -174 dBm represent?
A. The minimum detectable signal as a function of receive frequency
B. The theoretical noise in a 1 Hz bandwidth at the input of a perfect receiver at room temperature
C. The noise figure of a 1 Hz bandwidth receiver
D. The galactic noise contribution to minimum detectable signal

The value $-174\,$dBm is the theoretical noise in a 1-Hz bandwidth at the input of a perfect receiver at room temperature, as in **Answer B**.

E4C06 A CW receiver with the AGC off has an equivalent input noise power density of -174 dBm/Hz. What would be the level of an unmodulated carrier input to this receiver that would yield an audio output SNR of 0 dB in a 400 Hz noise bandwidth?
A. -174 dBm
B. -164 dBm
C. -155 dBm
D. -148 dBm

To find the theoretical signal strength that matches the noise level in the $400\,$Hz bandwidth at a SNR of $0\,$dB, we need to find the received signal power that matches the noise floor. The noise power, N, that the received signal matches is

$$N(dBm) = -174\,\text{dBm}/\text{Hz} + 10\log(400\,\text{Hz}) = -148\,\text{dBm}$$

Answer D has the correct computation result.

E4C07 What does the MDS of a receiver represent?
A. The meter display sensitivity
B. The minimum discernible signal
C. The multiplex distortion stability
D. The maximum detectable spectrum

A working definition of the Minimum Discernible Signal (MDS) is the minimum value of a signal that the system or sensor can detect. This signal level is also known as the system's *threshold*. **Answer B** is the correct choice. The other options are distractions.

E4C08 An SDR receiver is overloaded when input signals exceed what level?
A. One-half the maximum sample rate
B. One-half the maximum sampling buffer size
C. The maximum count value of the analog-to-digital converter
D. The reference voltage of the analog-to-digital converter

A Software Defined Radio (SDR) relies on an ADC, so if the input signal exceeds the ADC range, the SDR is overloaded. The ADC converter in Figure 4.2(b) is typically a "flash" ADC that has a limited input range controlled by a reference voltage. If the input signal exceeds that range, then the ADC is overloaded, and **Answer D** is the right choice. Answer C will result from this condition, but it is not the cause.

E4C09 Which of the following choices is a good reason for selecting a high frequency for the design of the IF in a superheterodyne HF or VHF communications receiver?
A. Fewer components in the receiver
B. Reduced drift
C. Easier for front-end circuitry to eliminate image responses
D. Improved receiver noise figure

If the Intermediate Frequency (IF) frequency is higher, then image frequencies lie at a farther distance from the carrier than with a lower frequency IF. This separation will make it easier for the front-end circuitry to filter out the potential images, as in **Answer C**. The other options are not electrically correct statements.

E4C10 What is an advantage of having a variety of receiver IF bandwidths from which to select?
A. The noise figure of the RF amplifier can be adjusted to match the modulation type, thus increasing receiver sensitivity
B. Receiver power consumption can be reduced when wider bandwidth is not required
C. Receive bandwidth can be set to match the modulation bandwidth, maximizing signal-to-noise ratio and minimizing interference
D. Multiple frequencies can be received simultaneously if desired

As we have seen, various modulation modes and data types have different transmission bandwidths. Also, the receiver noise is proportional to the IF bandwidth. Therefore, to minimize noise and have the best quality signal reception, we wish to have an IF bandwidth that is no smaller than the signal's bandwidth and, preferably, not much larger either. Maximizing the SNR of

Answer C shows the positive result of having selectable bandwidths that can match the signal.

E4C11 Why can an attenuator be used to reduce receiver overload on the lower frequency HF bands with little or no impact on signal-to-noise ratio?
  A. The attenuator has a low-pass filter to increase the strength of lower frequency signals
  B. The attenuator has a noise filter to suppress interference
  C. Signals are attenuated separately from the noise
  D. Atmospheric noise is generally greater than internally generated noise even after attenuation

In the lower High Frequency (HF) bands, atmospheric noise often plagues the receiver by swamping the signal and overloading the receiver. Operators can strategically place attenuators to reduce the input signal for the receiver to perform its signal processing functions and still produce an output of adequate quality. **Answer D** is the right choice in this situation.

E4C12 Which of the following has the largest effect on an SDR receiver's dynamic range?
  A. CPU register width in bits
  B. Anti-aliasing input filter bandwidth
  C. RAM speed used for data storage
  D. Analog-to-digital converter sample width in bits

The user determines the dynamic range by the maximum voltage range to be covered and the smallest voltage difference that the sampler can detect. In a SDR, the ADC sample width in bits controls the minimum voltage difference that the sampler can detect so that it will have the most significant effect on the dynamic range. This property makes **Answer D** the correct choice.

E4C13 How does a narrow-band roofing filter affect receiver performance?
  A. It improves sensitivity by reducing front end noise
  B. It improves intelligibility by using low Q circuitry to reduce ringing
  C. It improves dynamic range by attenuating strong signals near the receive frequency
  D. All of these choices are correct

The purpose of the roofing filter is to filter out strong signals that could interfere with the desired signal, as indicated in **Answer C**. The other options are distractions.

E4C14 What transmit frequency might generate an image response signal in a receiver tuned to 14.300 MHz and which uses a 455 kHz IF frequency?
- A. 13.845 MHz
- B. 14.755 MHz
- C. 14.445 MHz
- D. 15.210 MHz

As we saw in the General Class study guide, the image frequencies are on either side of the carrier and displaced by twice the IF frequency. The images will be at 15.210 MHz and 13.390 MHz. **Answer D** is the right choice. The other options do not use twice the IF frequency.

E4C15 What is reciprocal mixing?
- A. Two out-of-band signals mixing to generate an in-band spurious signal
- B. In-phase signals cancelling in a mixer resulting in loss of receiver sensitivity
- C. Two digital signals combining from alternate time slots
- D. Local oscillator phase noise mixing with adjacent strong signals to create interference to desired signals

As we saw above, phase noise and strong adjacent signals can cause problems in a receiver. The process of mixing the phase noise with the adjacent signal to create interference for the desired signal is reciprocal mixing as in **Answer D**.

4.6   E4D - Receiver Hardware Performance

4.6.1   Overview

The *Receiver Hardware Performance* question group in Subelement E4 further tests you on the characteristics of receiver operations. The *Receiver Hardware Performance* group covers topics such as
- Blocking dynamic range
- Intermodulation and cross-modulation interference
- Third-order intercept
- Desensitization
- Preselector

The test producer will select one of the 12 questions in this group for your exam.

4.6.2 Questions

E4D01 What is meant by the blocking dynamic range of a receiver?
  A. The difference in dB between the noise floor and the level of an incoming signal which will cause 1 dB of gain compression
  B. The minimum difference in dB between the levels of two FM signals which will cause one signal to block the other
  C. The difference in dB between the noise floor and the third-order intercept point
  D. The minimum difference in dB between two signals which produce third-order intermodulation products greater than the noise floor

As we saw earlier, strong adjacent signals can cause problems, including blocking signal reception. This problem is especially true for weak signals. The receiver's Blocking Dynamic Range (BDR) assesses how well this weak signal will pass through the receiver. A gain compression of 1 dB is the metric for not significantly interfering, and **Answer A** matches the standard definition.

E4D02 Which of the following describes problems caused by poor dynamic range in a receiver?
  A. Spurious signals caused by cross-modulation and desensitization from strong adjacent signals
  B. Oscillator instability requiring frequent retuning and loss of ability to recover the opposite sideband
  C. Cross-modulation of the desired signal and insufficient audio power to operate the speaker
  D. Oscillator instability and severe audio distortion of all but the strongest received signals

Looking at the previous questions, we see that the dynamic range is related to receiver sensitivity and noise floor levels. These do not affect the receiver oscillator so we can eliminate Answers B and D. Answer C is incorrect because the lack of dynamic range will not directly affect audio output power, so this is not a good choice. The right choice is **Answer A** since desensitization and spurious signals from adjacent signals would be a potential problem.

E4D03 How can intermodulation interference between two repeaters occur?
  A. When the repeaters are in close proximity and the signals cause feedback in the final amplifier of one or both transmitters
  B. When the repeaters are in close proximity and the signals mix in the final amplifier of one or both transmitters
  C. When the signals from the transmitters are reflected out of phase from airplanes passing overhead
  D. When the signals from the transmitters are reflected in phase from airplanes passing overhead

Reflecting signals off airplanes is a form of multipath propagation when the geometry is correct, and we can eliminate Answers C and D from consideration. The Intermodulation Distortion (IMD) interference is a mixing process, so **Answer B** has the correct definition. Answer A is close, but it uses feedback and not mixing as in the definition, so it is incorrect.

E4D04 Which of the following may reduce or eliminate intermodulation interference in a repeater caused by another transmitter operating in close proximity?
  A. A band-pass filter in the feed line between the transmitter and receiver
  B. A properly terminated circulator at the output of the repeater's transmitter
  C. Utilizing a Class C final amplifier
  D. Utilizing a Class D final amplifier

One may naturally think that installing some type of filter, as in Answer A, cures the problem. However, it will not work here because the "offending" transmitter will produce a solid signal in the desired passband, and the filter does not eliminate the interference. The designer can use a RF circulator element to route interfering signals away before they enter the receiver, and **Answer B** is the correct remedy among the choices given to answer this question. Amplifiers will only make things worse.

E4D05 What transmitter frequencies would cause an intermodulation-product signal in a receiver tuned to 146.70 MHz when a nearby station transmits on 146.52 MHz?
  A. 146.34 MHz and 146.61 MHz
  B. 146.88 MHz and 146.34 MHz
  C. 146.10 MHz and 147.30 MHz
  D. 173.35 MHz and 139.40 MHz

The question is asking for the IMD frequency that will show up at the receiver tuning frequency. The strongest IMD frequency components arise from $f_{IMD} = 2f_1 \pm f_2$ and $f_{IMD} = 2f_2 \pm f_1$. Here, f_1 is 146.52 MHz and f_{IMD} is 146.70 MHz. In this case, the equations we need are: $f_{IMD} = 2f_1 - f_2$ and $f_{IMD} = 2f_2 - f_1$. Using the first of these, $f_2 = 2f_1 - f_{IMD} = 2 \times 146.52\,\text{MHz} - 146.70\,\text{MHz} = 146.34\,\text{MHz}$, which we find in Answers A and B, so we can eliminate Answers C and D from further consideration. Using the second equation and some algebra, we have $f_2 = (f1 + f_{IMD})/2 = (146.52\,\text{MHz} + 146.70\,\text{MHz})/2 = 146.61\,\text{MHz}$, which we find in **Answer A**, so that is the correct choice.

E4D06 What is the term for spurious signals generated by the combination of two or more signals in a non-linear device or circuit?
  A. Amplifier desensitization
  B. Neutralization
  C. Adjacent channel interference
  D. Intermodulation

Based on the previous questions, you should be able to spot IMD, as in **Answer D**, as the right choice.

E4D07 Which of the following reduces the likelihood of receiver desensitization?
A. Decrease the RF bandwidth of the receiver
B. Raise the receiver IF frequency
C. Increase the receiver front end gain
D. Switch from fast AGC to slow AGC

As we have seen, desensitization occurs when a strong, unwanted signal is near the desired signal. An effective way to remove the unwanted signal is to reduce the RF bandwidth of the receiver so that the receiver no longer sees the unwanted signal. **Answer A** matches this method.

E4D08 What causes intermodulation in an electronic circuit?
A. Too little gain
B. Lack of neutralization
C. Nonlinear circuits or devices
D. Positive feedback

The choices here are not very specific, and the most that we can say is that nonlinear effects cause the IMD, so **Answer C** is the best choice to answer the question. Too little gain is not the issue, so Answer A is incorrect. Neutralization and positive feedback affect circuits but do not cause IMD by themselves.

E4D09 What is the purpose of the preselector in a communications receiver?
A. To store often-used frequencies
B. To provide a range of AGC time constants
C. To increase rejection of signals outside the desired band
D. To allow selection of the optimum RF amplifier device

The preselector will increase sensitivity and assist in rejecting unwanted signals outside of the desired band, so **Answer C** is the best choice to answer this question. The other options are nice for a receiver but unrelated to a preselector's function.

E4D10 What does a third-order intercept level of 40 dBm mean with respect to receiver performance?
- A. Signals less than 40 dBm will not generate audible third-order intermodulation products
- B. The receiver can tolerate signals up to 40 dB above the noise floor without producing third-order intermodulation products
- C. A pair of 40 dBm input signals will theoretically generate a third-order intermodulation product that has the same output amplitude as either of the input signals
- D. A pair of 1 mW input signals will produce a third-order intermodulation product which is 40 dB stronger than the input signal

Answer C has the correct interpretation: a pair of 40 dBm signals will theoretically generate the same output on the third-order intermodulation frequency as on the input frequency.

E4D11 Why are odd-order intermodulation products, created within a receiver, of particular interest compared to other products?
- A. Odd-order products of two signals in the band of interest are also likely to be within the band
- B. Odd-order products overload the IF filters
- C. Odd-order products are an indication of poor image rejection
- D. Odd-order intermodulation produces three products for every input signal within the band of interest

As we saw earlier, the thing to remember is that the odd-order intermodulation products may be present within the band of interest. This result means you should choose **Answer A** to answer this question.

E4D12 What is the term for the reduction in receiver sensitivity caused by a strong signal near the received frequency?
- A. Desensitization
- B. Quieting
- C. Cross-modulation interference
- D. Squelch gain rollback

Desensitization is the reduction in receiver sensitivity by nearby signals, so **Answer A** is the right choice for this question. Quieting is a good thing with FM receivers since it indicates that the FM receiver has locked onto the FM carrier; however, it does not apply in this instance. Answer C is a bad problem with receiving another, unwanted signal, but it is not a reduction in receiver sensitivity, so this is not the correct answer either. Answer D is technobabble.

4.7 E4E - Noise Suppression and Interference

4.7.1 Overview

The *Noise Suppression and Interference* question group in Subelement E4 quizzes you on the characteristics of receiver noise. The *Noise Suppression and Interference* group covers topics such as
- System noise
- Electrical appliance noise
- Line noise
- Locating noise sources
- DSP noise reduction
- Noise blankers
- Grounding for signals
- Common mode currents

The test producer will select one of the 11 questions in this group for your exam.

4.7.2 Questions

E4E01 What problem can occur when using an automatic notch filter (ANF) to remove interfering carriers while receiving CW signals?
- A. Removal of the CW signal as well as the interfering carrier
- B. Any nearby signal passing through the DSP system will overwhelm the desired signal
- C. Received CW signals will appear to be modulated at the DSP clock frequency
- D. Ringing in the DSP filter will completely remove the spaces between the CW characters

A notch filter removes the components in a narrow band around a specific frequency. If the notch filter removes the interfering signals, it may also remove desired signals because a filter cannot be infinitely sharp. **Answer A** matches this reasoning.

E4E02 Which of the following types of noise can often be reduced with a digital signal processing noise filter?
- A. Broadband white noise
- B. Ignition noise
- C. Power line noise
- D. All of these choices are correct

Engineers can design Digital Signal Processor (DSP) noise filters for each noise type listed in Answers A, B, and C, so **Answer D** is the best choice for this question.

E4E03 Which of the following signals might a receiver noise blanker be able to remove from desired signals?
   A. Signals that are constant at all IF levels
   B. Signals that appear across a wide bandwidth
   C. Signals that appear at one IF but not another
   D. Signals that have a sharply peaked frequency distribution

A receiver noise blanker can remove signals correlated across a wide bandwidth, so **Answer B** is the correct choice.

E4E04 How can conducted and radiated noise caused by an automobile alternator be suppressed?
   A. By installing filter capacitors in series with the DC power lead and a blocking capacitor in the field lead
   B. By installing a noise suppression resistor and a blocking capacitor in both leads
   C. By installing a high-pass filter in series with the radio's power lead and a low-pass filter in parallel with the field lead
   D. By connecting the radio's power leads directly to the battery and by installing coaxial capacitors in line with the alternator leads

Answer D gives the correct technique, so that is the right choice for this question. Answer A is not a good choice because the radio does not have a "field lead." Answer B is wrong because a resistor in the power lead will cause problems in power transfer. Answer C is not a good choice since a high pass filter will pass more noise.

E4E05 How can radio frequency interference from an AC motor be suppressed?
   A. By installing a high-pass filter in series with the motor's power leads
   B. By installing a brute-force AC-line filter in series with the motor leads
   C. By installing a bypass capacitor in series with the motor leads
   D. By using a ground-fault current interrupter in the circuit used to power the motor

The Ground Fault Circuit Interrupter (GFCI) in Answer D will help with electrical shorts, but not electrical noise, so this is not a good choice for this question. A capacitor in series will not filter noise, so Answer C is not a good choice. Noise has high-frequency components, and Answer A will pass them. The Alternating Current (AC) line filter in **Answer B** is the best choice among those given for this question.

E4E06 What is one type of electrical interference that might be caused by a nearby personal computer?
A. A loud AC hum in the audio output of your station receiver
B. A clicking noise at intervals of a few seconds
C. The appearance of unstable modulated or unmodulated signals at specific frequencies
D. A whining type noise that continually pulses off and on

If you place an Amplitude Modulation (AM) receiver near a computer, you will hear the noise being emitted by the computer while it is on. Of the choices given, **Answer C** is the best description of the electronic noise that the computer generates while operating.

E4E07 Which of the following can cause shielded cables to radiate or receive interference?
A. Low inductance ground connections at both ends of the shield
B. Common-mode currents on the shield and conductors
C. Use of braided shielding material
D. Tying all ground connections to a common point resulting in differential-mode currents in the shield

Common-mode currents, as in **Answer B**, are frequent causes of spurious signals in communications systems. The other choices are good items for the grounding system and help remove noise.

E4E08 What current flows equally on all conductors of an unshielded multi-conductor cable?
A. Differential-mode current
B. Common-mode current
C. Reactive current only
D. Return current

This question defines common-mode current, as in **Answer B**. Differential-mode current occurs on parallel wires, so it is incorrect here. The return current is the current back to the energy source. Reactive current is in a capacitor or an inductor.

E4E09 What undesirable effect can occur when using an IF noise blanker?
A. Received audio in the speech range might have an echo effect
B. The audio frequency bandwidth of the received signal might be compressed
C. Nearby signals may appear to be excessively wide even if they meet emission standards
D. FM signals can no longer be demodulated

Because of the correlation properties of many noise blankers, they can make the signals appear wider. This property makes **Answer C** the correct choice.

E4E10 What might be the cause of a loud roaring or buzzing AC line interference that comes and goes at intervals?
  A. Arcing contacts in a thermostatically controlled device
  B. A defective doorbell or doorbell transformer inside a nearby residence
  C. A malfunctioning illuminated advertising display
  D. All of these choices are correct

Each choice given in Answers A, B, and C is a possibility, which makes **Answer D** the best answer for this question.

E4E11 What could cause local AM broadcast band signals to combine to generate spurious signals in the MF or HF bands?
  A. One or more of the broadcast stations is transmitting an over-modulated signal
  B. Nearby corroded metal joints are mixing and re-radiating the broadcast signals
  C. You are receiving sky wave signals from a distant station
  D. Your station receiver IF amplifier stage is defective

Since the AM band does not overlap the Amateur Service bands, there needs to be some object mixing the AM frequencies with another signal and having the resulting signal appear in the amateur band. **Answer B**, even though it sounds improbable, is the correct choice here.

Chapter 5

E5 — ELECTRICAL PRINCIPLES

5.1 Introduction

The *Electrical Principals* subelement will quiz you on more advanced circuit analysis than what we saw with Ohm's Law. Here, we will see questions on Alternating Current (AC) circuit resonance, bandwidth, and time constants. We will also see complex numbers and the phasor representation for typical circuit elements' voltage, current, and reactance. We will use the phasor representation with AC circuit power computations. The *Electrical Principals* subelement has the following question groups:

- A. Resonance and Q
- B. Time constants and phase relationships
- C. Coordinate systems and phasors in electronics
- D. AC and RF energy in real circuits

Subelement 5 will generate four questions on the Extra examination.

5.2 Radio Engineering Concepts

Resonant Circuits and Quality Factor Resonant circuits are special cases of the general Resistor-Inductor-Capacitor (RLC) circuit configurations shown in Figure 5.1. When a circuit is in resonance, the inductive and capacitive reactances cancel, and only the resistive element remains. When in resonance, the voltage and current are in phase because the reactive elements have canceled. Because the reactances are frequency dependent, the resonance condition is only for a specific frequency. The circuit will show a net inductive or capacitive reactance at other frequencies.

One measure of how tightly the resonance region is in the circuit is by use

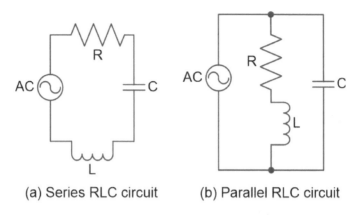

(a) Series RLC circuit   (b) Parallel RLC circuit

Figure 5.1: Simple series and parallel RLC circuit configurations.

of the Quality Factor (Q). A small Q indicates that the circuit is not highly selective, and engineers say it has a broad bandwidth. A large Q indicates that the circuit is very selective and has a narrow bandwidth. Because we know that the reactive components cancel in Figure 5.1, we can write

$$X_L = 2\pi f L = X_C = \frac{1}{2\pi f C}$$

We can use this relationship to solve for the resonant frequency, f_r

$$f_r = \frac{1}{2\pi\sqrt{LC}}$$

Circuit analysts compute the Q for series circuits in terms of the total reactance, X_T, and the total resistance, R_T, using

$$Q = \frac{X_T}{R_T}$$

Notice that Q is unitless because both the reactance and resistance have the units of ohms. For Q larger than 10, we estimate the 3-dB bandwidth, B, for the circuit at the resonant frequency, f_r, from

$$B = \frac{f_r}{Q}$$

For parallel circuits, we find the Q from

$$Q = \frac{R_T}{X_T}$$

Charging

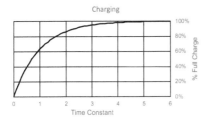

Discharging

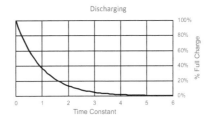

(a) Charging as a function of the time constant.

(b) Discharging as a function of the time constant.

Figure 5.2: Reactive circuit charge and discharge timing as a function of the circuit's time constant.

Time Constants When reactive components are either energized or dissipate their energy, it does not happen instantaneously. Rather, the charge or discharge time follows and exponential curve. The time constant, τ, sets the pace for the cycle. Figure 5.2 illustrates the process. The charging voltage level, V_c, in terms of 100 % of the full charge is

$$V_c = \left(1 - e^{-t/\tau}\right) \times 100\,\%$$

The discharging voltage level, V_d, in terms of 100 % of the full charge is

$$V_d = \left(e^{-t/\tau}\right) \times 100\,\%$$

As a rule of thumb, we can assume that the charge or discharge process is complete after five time constants. For a resistor-capacitor circuit, $\tau = RC$, while for a resistor-inductor circuit, $\tau = L/R$.

Phasors In circuit analysis, we learn that we can represent the reactance for capacitors, X_C, and inductors, X_L, in terms of the frequency, f, as
Capacitive Reactance $X_C = -j/\left(2\pi fC\right)$
Inductive Reactance $X_L = j2\pi fL$
The ideal resistor has the same resistance, R, for all frequencies. The resistance or reactance unit for all RLC components is the ohm (Ω). The total impedance, Z, in ohms is then $Z = R_T + jX_T$ with R_T representing the total resistance and X_T the total reactance.

The admittance, Y, is related to the impedance by $Y = 1/Z$. The admittance is in units of Siemens (S). We can write the admittance like the impedance using $Y = G + jB$ where G is the conductance, and B is the susceptance, both also having the units of siemens.

We can apply Ohm's Law for each element individually. We compute voltage

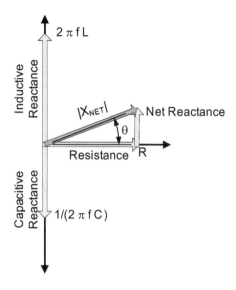

Figure 5.3: Rectangular and polar representations for RLC circuit elements.

across each element using

$$V_R = RI_R$$
$$V_L = X_L I_L$$
$$V_C = X_C I_C$$

Suppose we have a RLC circuit like the one shown in Figure 5.1(a). Figure 5.3 shows how we place each circuit element on a rectangular axis. We plot the resistor values along the $+x$ axis, the inductor values along the $+y$ axis, and the capacitor values along the $-y$ axis. We can combine the two reactances on the plot to form a net reactance shown by the smaller vector. If we apply Ohm's Law to this circuit, we have $V = ZI$ where $Z = R + j2\pi fL - j/(2\pi fC)$ or $Z = R + jX_{NET}$.

For complex number representations, we can convert between the rectangular and the polar form by using the real and imaginary parts (the resistance part and the reactance part without the "j") to form the magnitude and phase angle:

$$|Z| = \sqrt{R^2 + X^2}$$
$$\theta = tan^{-1}(X/R)$$

Which we write as $|Z| \angle \theta$. This is the phasor representation for the quantity.

When we are using AC sources and signals, we find it easier to perform the circuit analysis using the complex number representation and use the phasor representation for the voltage, current, and reactance. The phasor forms for the RLC components are

Resistor $X_R = R \angle 0°$
Capacitor $X_C = 1/(2\pi f C) \angle - 90°$
Inductor $X_L = 2\pi f L \angle 90°$
We can now use this in Ohm's Law to find the voltage across each component using

$$| V_R | \angle\theta = R | I_R | \angle\theta$$
$$| V_L | \angle\theta + 90° = | X_L | \angle 90° | I_L | \angle\theta$$
$$| V_C | \angle\theta - 90° = | X_C | \angle - 90° | I_C | \angle\theta$$

Notice that for capacitors inductors, the voltage across and the current through the elements are not in phase but differ by 90°. Engineers call this the voltage leading the current in the inductor (the voltage's phase is 90° ahead of the current's phase) and the voltage lagging the current in the capacitor (the voltage's phase is 90° behind the current's phase). Also, notice that it is equally valid to say that the current lags the voltage in the inductor and leads the voltage in the capacitor.

Power and Power Factor With inductors and capacitors in the circuit, we need to be careful of how we compute power. The power computation needs to account for the reactive elements storing energy in their fields. The circuit does not lose the energy stored in the reactive elements as it does for the energy delivered to the resistor that dissipates the energy as heat. In practice, we define two types of power

Real Power is the power delivered to the resistive part of the circuit; has the units of watts (W)

Apparent Power is the product of the current through and the voltage across an impedance; has the units of volt-amperes (VA)

Engineers call the ratio of the real power to the apparent power the Power Factor (PF). The PF is also, equivalently, defined in terms of the circuit's total impedance. In that case, $PF = \cos\theta$ where $\theta = \arctan X/R$. After the power factor, one also adds "leading" or "lagging" to specify the voltage's phase relative to the current.

5.3 E5A - Resonance and Q

5.3.1 Overview

The *Resonance and Q* question group in Subelement E5 tests you on the characteristics of resonant circuits and their quality factor, Q. The *Resonance and Q* group covers topics such as

- Characteristics series and parallel resonant circuits
- Definitions and effects of Q; half-power bandwidth
- Phase relationships in reactive circuits

The test producer will select one of the 16 questions in this group for your exam.

5.3.2   Questions

E5A01 What can cause the voltage across reactances in a series RLC circuit to be higher than the voltage applied to the entire circuit?
- A. Resonance
- B. Capacitance
- C. Conductance
- D. Resistance

Resistance, capacitance, and conductance are bulk properties of materials, and, individually, they do not cause the effect described in the question, so Answers B, C, and D are incorrect. Resonance is an interaction between components, and it can cause an increase in voltage, so **Answer A** is the correct choice.

E5A02 What is resonance in an LC or RLC circuit?
- A. The highest frequency that will pass current
- B. The lowest frequency that will pass current
- C. The frequency at which the capacitive reactance equals the inductive reactance
- D. The frequency at which the reactive impedance equals the resistive impedance

An electrical circuit will pass at least a small amount of current at each frequency, so Answers A and B are not good choices. **Answer C** is the correct answer because it describes the mutual cancellation of the capacitive and inductive components. Answer D does not use the terms correctly.

E5A03 What is the magnitude of the impedance of a series RLC circuit at resonance?
- A. High, as compared to the circuit resistance
- B. Approximately equal to capacitive reactance
- C. Approximately equal to inductive reactance
- D. Approximately equal to circuit resistance

Next, we have several questions about order-of-magnitude estimates of circuit parameters. In a resonant circuit, the L and C parts mostly cancel while the R remains making **Answer D** the correct choice. Since the resonance removes the L and C components, we can eliminate Answers B and C. Answer A is a distraction.

E5A04 What is the magnitude of the impedance of a parallel RLC circuit at resonance?
- A. Approximately equal to circuit resistance
- B. Approximately equal to inductive reactance
- C. Low, as compared to the circuit resistance
- D. High compared to the circuit resistance

Just because we have changed to a parallel circuit, the basic principle does not change. As with series resonance, a parallel resonant circuit has the L and C parts mostly cancel while the R remains, making **Answer A** the correct choice. Since the resonance removes the L and C components, we eliminate Answers B and D. Answer C is a distraction.

E5A05 What is the result of increasing the Q of an impedance-matching circuit?
- A. Matching bandwidth is decreased
- B. Matching bandwidth is increased
- C. Matching range is increased
- D. It has no effect on impedance matching

A high Q implies a narrower bandwidth in the circuit, so **Answer A** is the right choice. Be careful with Answer B because it is the opposite effect.

E5A06 What is the magnitude of the circulating current within the components of a parallel LC circuit at resonance?
- A. It is at a minimum
- B. It is at a maximum
- C. It equals 1 divided by the quantity 2 times pi, multiplied by the square root of inductance L multiplied by capacitance C
- D. It equals 2 multiplied by pi, multiplied by frequency, multiplied by inductance

This question may be another one for which you may need to memorize the correct answer. You need to remember that the circulating current is maximum for the parallel LC case, so **Answer B** is the correct choice. Answer A is to confuse you.

E5A07 What is the magnitude of the current at the input of a parallel RLC circuit at resonance?
- A. Minimum
- B. Maximum
- C. R/L
- D. L/R

Here you need to know that the input current is minimum for the parallel RLC circuit, so **Answer A** is the correct choice. Answer B is to distract you.

E5A08 What is the phase relationship between the current through and the voltage across a series resonant circuit at resonance?
   A. The voltage leads the current by 90 degrees
   B. The current leads the voltage by 90 degrees
   C. The voltage and current are in phase
   D. The voltage and current are 180 degrees out of phase

At resonance, the current and voltage are in phase, so **Answer C** is the correct choice. The other options are incorrect for a resonant circuit.

E5A09 How is the Q of an RLC parallel resonant circuit calculated?
   A. Reactance of either the inductance or capacitance divided by the resistance
   B. Reactance of either the inductance or capacitance multiplied by the resistance
   C. Resistance divided by the reactance of either the inductance or capacitance
   D. Reactance of the inductance multiplied by the reactance of the capacitance

Here we have two questions with answers that appear very similar and are easy to misread. The Q for a parallel resonant circuit is the ratio of the parallel resistance to the parallel reactance as in **Answer C**. Answer A is for the series resonant circuit, so be careful when reading the answers.

E5A10 How is the Q of an RLC series resonant circuit calculated?
   A. Reactance of either the inductance or capacitance divided by the resistance
   B. Reactance of either the inductance or capacitance multiplied by the resistance
   C. Resistance divided by the reactance of either the inductance or capacitance
   D. Reactance of the inductance multiplied by the reactance of the capacitance

The Q for a series resonant circuit is the ratio of the series reactance to the series resistance as in **Answer A**. Answer C is for the parallel resonant circuit, so be careful when reading the answers.

E5A11 What is the half-power bandwidth of a resonant circuit that has a resonant frequency of 7.1 MHz and a Q of 150?
   A. 157.8 Hz
   B. 315.6 Hz
   C. 47.3 kHz
   D. 23.67 kHz

The relationship between bandwidth, B, resonant frequency, f_r, and quality, Q, is $B = f_r/Q$. Using the numbers given, $B = 7.1\,\text{MHz}/150 = 47.3\,\text{kHz}$, so **Answer C** is the correct choice. The other answers have math errors.

E5A12 What is the half-power bandwidth of a resonant circuit that has a resonant frequency of 3.7 MHz and a Q of 118?
   A. 436.6 kHz
   B. 218.3 kHz
   C. 31.4 kHz
   D. 15.7 kHz

Here we use the same relationship as the previous question to solve for the bandwidth. Using the numbers given, $B = 3.7\,\text{MHz}/118 = 31.4\,\text{kHz}$, so **Answer C** is the correct choice. The other choices have distraction math errors.

E5A13 What is an effect of increasing Q in a series resonant circuit?
   A. Fewer components are needed for the same performance
   B. Parasitic effects are minimized
   C. Internal voltages increase
   D. Phase shift can become uncontrolled

Generally, increasing the Q will not result in fewer components or minimizing parasitic effects, so those are incorrect choices. However, the internal voltages increase, so **Answer C** is the correct choice.

E5A14 What is the resonant frequency of an RLC circuit if R is 22 ohms, L is 50 microhenrys and C is 40 picofarads?
   A. 44.72 MHz
   B. 22.36 MHz
   C. 3.56 MHz
   D. 1.78 MHz

Remember, resonance is controlled by the reactive components. From circuit theory, we learn that RLC circuits have a resonant frequency at $f_r = 1/2\pi\sqrt{LC}$. For this circuit, $f_r = 1/2\pi\sqrt{50\,\mu\text{H}40\,\text{pF}} = 3.56\,\text{MHz}$. **Answer C** is correct, while the other choices contain math errors.

E5A15 Which of the following increases Q for inductors and capacitors?
   A. Lower losses
   B. Lower reactance
   C. Lower self-resonant frequency
   D. Higher self-resonant frequency

Lowering the losses in the circuit will increase the Q, so **Answer A** is the correct choice.

E5A16 What is the resonant frequency of an RLC circuit if R is 33 ohms, L is 50 microhenrys and C is 10 picofarads?
  A. 23.5 MHz
  B. 23.5 kHz
  C. 7.12 kHz
  D. 7.12 MHz

Using the same circuit theory, $f_r = 1/2\pi\sqrt{50\,\mu H 10\,pF} = 7.12\,MHz$. **Answer D** is correct, while the other choices have math errors. Be careful with Answer C because it has the correct numbers, but kHz replaces the MHz unit.

5.4   E5B - Time Constants and Phase Relationships

5.4.1   Overview

The *Time Constants and Phase Relationships* question group in Subelement E5 tests you on phase relationships for circuit elements and circuit time constants. The *Time Constants and Phase Relationships* group covers topics such as
  • RL and RC time constants
  • Phase angle in reactive circuits and components
  • Admittance and susceptance
The test producer will select one of the 12 questions in this group for your exam.

5.4.2   Questions

E5B01 What is the term for the time required for the capacitor in an RC circuit to be charged to 63.2 % of the applied voltage or to discharge to 36.8 % of its initial voltage?
  A. An exponential rate of one
  B. One time constant
  C. One exponential period
  D. A time factor of one

The question defines reaching the first time constant in the charging or discharging time, so **Answer B** is the correct choice.

E5B02 What letter is commonly used to represent susceptance?
  A. G
  B. X
  C. Y
  D. B

The usual symbol for the susceptance is the letter B making **Answer D** the right choice. G is the conductance, X is the reactance, and Y is the admittance.

Remember: $Y = 1/Z = G + jB$ and $Z = R + jX$.

E5B03 How is impedance in polar form converted to an equivalent admittance?
A. Take the reciprocal of the angle and change the sign of the magnitude
B. Take the reciprocal of the magnitude and change the sign of the angle
C. Take the square root of the magnitude and add 180 degrees to the angle
D. Square the magnitude and subtract 90 degrees from the angle

Because impedance and admittance are reciprocal quantities, we apply the mathematical rule for finding reciprocals in polar form: take the reciprocal of the magnitude and invert the sign of the angle. **Answer B** is the correct method, and the other answers represent math mistakes.

E5B04 What is the time constant of a circuit having two 220-microfarad capacitors and two 1-megohm resistors, all in parallel?
A. 55 seconds
B. 110 seconds
C. 440 seconds
D. 220 seconds

In this circuit, the total resistance is $500\,\text{k}\Omega$. The total capacitance is $440\,\mu\text{F}$. The time constant is $\tau = RC = 500\,\text{k}\Omega \times 440\,\mu\text{F} = 220\,\text{s}$. **Answer D** gives the correct answer, and the others contain math mistakes.

E5B05 What happens to the magnitude of a pure reactance when it is converted to a susceptance?
A. It is unchanged
B. The sign is reversed
C. It is shifted by 90 degrees
D. It becomes the reciprocal

As we mentioned above, the magnitude becomes the reciprocal of the original value, and **Answer D** is the correct choice. Answer B is for the phase angle, so it is incorrect here.

E5B06 What is susceptance?
A. The magnetic impedance of a circuit
B. The ratio of magnetic field to electric field
C. The imaginary part of admittance
D. A measure of the efficiency of a transformer

By now, you should be able to spot that the susceptance, B, is the imaginary part of admittance ($Y = G + jB$), making **Answer C** the correct choice. The other options are to distract you.

E5B07 What is the phase angle between the voltage across and the current through a series RLC circuit if XC is 500 ohms, R is 1 kilohm, and XL is 250 ohms?
   A. 68.2 degrees with the voltage leading the current
   B. 14.0 degrees with the voltage leading the current
   C. 14.0 degrees with the voltage lagging the current
   D. 68.2 degrees with the voltage lagging the current

The total reactance in this series configuration is $250\,\Omega - 500\,\Omega$. Since the net reactance is negative or lagging, the phase angle in the answer needs to be negative, so Answers A and B are incorrect. Because the net reactance is smaller than the resistance, the phase angle needs to be less than $45°$ in magnitude, so Answer C seems to be the best choice. The phase angle between the voltage and the current is $\tan^{-1}(-250\,\Omega/1000\,\Omega) = -14.0°$, which corresponds to **Answer C**. Answer B has a leading phase angle, so read the question carefully.

E5B08 What is the phase angle between the voltage across and the current through a series RLC circuit if XC is 100 ohms, R is 100 ohms, and XL is 75 ohms?
   A. 14 degrees with the voltage lagging the current
   B. 14 degrees with the voltage leading the current
   C. 76 degrees with the voltage leading the current
   D. 76 degrees with the voltage lagging the current

Here we have another example of this type of analysis. The total reactance in this series configuration is $75\,\Omega - 100\,\Omega$. Since the net reactance is negative or lagging, the phase angle in the answer needs to be negative, so Answers B and C are incorrect. Because the net reactance is smaller than the resistance, the phase angle needs to be less than $45°$ in magnitude, so Answer A is the best choice. The phase angle between the voltage and the current is $\tan^{-1}(-25\,\Omega/100\,\Omega) = -14.0°$, which corresponds to **Answer A**.

E5B09 What is the relationship between the AC current through a capacitor and the voltage across a capacitor?
   A. Voltage and current are in phase
   B. Voltage and current are 180 degrees out of phase
   C. Voltage leads current by 90 degrees
   D. Current leads voltage by 90 degrees

In a capacitor, the current leads the voltage by $90°$, so **Answer D** gives the correct response. Answer C corresponds to an inductor and not a capacitor, as the question asks for, so this is incorrect. Answer A is for a resistor, so this is incorrect as well. Answer B is not for a single R, L, or C.

E5B10 What is the relationship between the AC current through an inductor and the voltage across an inductor?
  A. Voltage leads current by 90 degrees
  B. Current leads voltage by 90 degrees
  C. Voltage and current are 180 degrees out of phase
  D. Voltage and current are in phase

This question is a complement to the previous one. In an inductor, the voltage leads the current by 90°, so **Answer A** gives the correct response. Answer B is for a capacitor, so this is incorrect. Answer D is for a resistor, so this is incorrect as well. Answer C is not for R, L, or C elements.

E5B11 What is the phase angle between the voltage across and the current through a series RLC circuit if XC is 25 ohms, R is 100 ohms, and XL is 50 ohms?
  A. 14 degrees with the voltage lagging the current
  B. 14 degrees with the voltage leading the current
  C. 76 degrees with the voltage lagging the current
  D. 76 degrees with the voltage leading the current

The total reactance in this series configuration is $50\,\Omega - 25\,\Omega$. Since the net reactance is positive or leading, the phase angle in the answer needs to be positive, eliminating Answers A and C. Because the net reactance is smaller than the resistance, the phase angle needs to be less than 45° in magnitude. The phase angle between the voltage and the current is $\tan^{-1}(25\,\Omega/100\,\Omega) = 14.0°$, which corresponds to **Answer B**.

E5B12 What is admittance?
  A. The inverse of impedance
  B. The term for the gain of a field effect transistor
  C. The turns ratio of a transformer
  D. The unit used for Q factor

The admittance is the inverse of the impedance ($Y = 1/Z$), making **Answer A** the correct choice. Answer D cannot be correct because Q is unitless. Answers B and C are electrical quantities not related to the definition of admittance.

5.5 E5C - Coordinate Systems and Phasors

5.5.1 Overview

The *Coordinate Systems and Phasors* question group in Subelement E5 tests you on using phasors to represent circuit quantities. The *Coordinate Systems and Phasors* group covers topics such as

- Rectangular Coordinates
- Polar Coordinates
- Phasors

The test producer will select one of the 12 questions in this group for your exam.

5.5.2   Questions

E5C01 Which of the following represents a capacitive reactance in rectangular notation?
  A. $-jX$
  B. $+jX$
  C. Delta
  D. Omega

We represent a capacitive reactance with $-jX$, as in **Answer A**. Be careful because Answer B is for an inductor. Omega (Ω) symbolizes ohms and not a polar or rectangular notation. Delta is to distract you.

E5C02 How are impedances described in polar coordinates?
  A. By X and R values
  B. By real and imaginary parts
  C. By phase angle and amplitude
  D. By Y and G values

Polar coordinates, such as in Figure 5.3, use a phase angle, θ, and amplitude as in **Answer C**. Answer B is for rectangular coordinates, so it has the wrong form. The other options are to distract you.

E5C03 Which of the following represents an inductive reactance in polar coordinates?
  A. A positive magnitude
  B. A negative magnitude
  C. A positive phase angle
  D. A negative phase angle

Inductive reactances have a positive phase angle making **Answer C** the correct response. A negative phase angle is for a capacitor, so it is incorrect. Resistors, capacitors, and inductors all have a positive magnitude making Answer A incorrect. *RLC* components do not have negative magnitudes.

E5C04 What coordinate system is often used to display the resistive, inductive, and/or capacitive reactance components of impedance?
A. Maidenhead grid
B. Faraday grid
C. Elliptical coordinates
D. Rectangular coordinates

In addition to polar coordinates, as in the previous questions, we can also use rectangular coordinates as in **Answer D**. A Maidenhead grid is for your location on the Earth's surface, so it will not help here. The other choices are to distract you.

E5C05 What is the name of the diagram used to show the phase relationship between impedances at a given frequency?
A. Venn diagram
B. Near field diagram
C. Phasor diagram
D. Far field diagram

The diagram is a phasor diagram, as in **Answer C**. The Venn diagram is for sets, while the field diagrams are for antennas, so both are incorrect choices.

E5C06 What does the impedance 50-j25 represent?
A. 50 ohms resistance in series with 25 ohms inductive reactance
B. 50 ohms resistance in series with 25 ohms capacitive reactance
C. 25 ohms resistance in series with 50 ohms inductive reactance
D. 25 ohms resistance in series with 50 ohms capacitive reactance

This corresponds to a 50-Ω resistance in series with a 25-Ω capacitive reactance as in **Answer B**. The other options have the components swapped or try to make you think this is an inductor.

E5C07 Where is the impedance of a pure resistance plotted on rectangular coordinates?
A. On the vertical axis
B. On a line through the origin, slanted at 45 degrees
C. On a horizontal line, offset vertically above the horizontal axis
D. On the horizontal axis

A pure resistance has no reactive part, so it lies entirely on the positive *x-axis* on a rectangular coordinate plot. **Answer D** is the correct choice in the case. Answer A is for either a pure capacitor or a pure inductor. Answers B and C are to distract you.

E5C08 What coordinate system is often used to display the phase angle of a circuit containing resistance, inductive and/or capacitive reactance?
  A. Maidenhead grid
  B. Faraday grid
  C. Elliptical coordinates
  D. Polar coordinates

By now, you should be able to spot polar coordinates as in **Answer D** as the correct choice. The others are to distract you.

E5C09 When using rectangular coordinates to graph the impedance of a circuit, what do the axes represent?
  A. The X axis represents the resistive component and the Y axis represents the reactive component
  B. The X axis represents the reactive component and the Y axis represents the resistive component
  C. The X axis represents the phase angle and the Y axis represents the magnitude
  D. The X axis represents the magnitude and the Y axis represents the phase angle

In rectangular coordinates, such as in Figure 5.3, the *x-axis* represents the resistive component while the reactive component is on the *y-axis*, as in **Answer A**. The other choices are mere distractions.

Figure E5-1

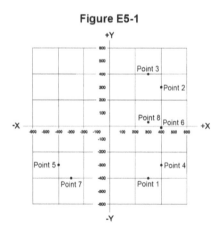

Figure 5.4: Question pool figure E5-1.

E5C10 Which point on Figure E5-1 best represents the impedance of a series circuit consisting of a 400-ohm resistor and a 38-picofarad capacitor at 14 MHz?
  A. Point 2
  B. Point 4
  C. Point 5
  D. Point 6

Some things to notice about this chart that will help with the questions in this section: (1) no actual *RLC* circuits will be found in quadrants II and III since regular passive resistors do not have a negative resistance; (2) purely resistive circuits will be found along the *x-axis*, purely inductive reactances will lie along the +*y-axis*, while purely capacitive reactances will lie along the -*y-axis*; (3) if the circuit is in resonance, it will lie along the *x-axis* since the inductive and capacitive reactances cancel out; (4) if the circuit has a net inductive reactance, it will lie in quadrant I, while if the circuit has a net capacitive reactance, it will lie in quadrant IV; (5) circuits in quadrant I have positive (leading) phase angles, while those in quadrant IV have negative (lagging) phase angles; (6) if the magnitude of the resistance is greater than the magnitude of the net reactance, the phase angle will be less than $45°$, while if the magnitude of the net reactance is greater than the magnitude of the resistance, the phase angle will be greater than $45°$.

Here, the total impedance is $Z = 400\,\Omega - j/[2\pi(14\,\text{MHz})(38\,\text{pF})] = 400\,\Omega - j299\,\Omega$ for this question. This impedance corresponds to Point 4 in quadrant IV on the chart, so **Answer B** is the right choice.

E5C11 Which point in Figure E5-1 best represents the impedance of a series circuit consisting of a 300-ohm resistor and an 18-microhenry inductor at 3.505 MHz?

A. Point 1
B. Point 3
C. Point 7
D. Point 8

The impedance of the circuit is $Z = 300\,\Omega + j2\pi(18\,\mu\text{H})(3.505\,\text{MHz}) = 300\,\Omega + j396.4\,\Omega$. This corresponds to Point 3 in quadrant I on the graph, so **Answer B** is the right choice.

E5C12 Which point on Figure E5-1 best represents the impedance of a series circuit consisting of a 300-ohm resistor and a 19-picofarad capacitor at 21.200 MHz?

A. Point 1
B. Point 3
C. Point 7
D. Point 8

Back to capacitors and we compute the impedance as we did earlier. $Z = 300\,\Omega - j/[2\pi(21.200\,\text{MHz})(19\,\text{pF})] = 300\,\Omega - j395\,\Omega$. This corresponds to Point 1 in quadrant IV on the chart, so **Answer A** is the right choice.

5.6 E5D - AC and RF Energy in Real Circuits

5.6.1 Overview

The *AC and RF Energy in Real Circuits* question group in Subelement E5 tests you on various physical effects that appear in circuits and wiring. We also examine power factor computations. The *AC and RF Energy in Real Circuits* group covers topics such as

- Skin effect
- Electromagnetic fields
- Reactive power
- Power factor
- Electrical length of conductors at UHF and microwave frequencies
- Microstrip

The test producer will select one of the 15 questions in this group for your exam.

5.6.2   Questions

E5D01 What is the result of skin effect?
  A. As frequency increases, RF current flows in a thinner layer of the conductor, closer to the surface
  B. As frequency decreases, RF current flows in a thinner layer of the conductor, closer to the surface
  C. Thermal effects on the surface of the conductor increase the impedance
  D. Thermal effects on the surface of the conductor decrease the impedance

The skin effect is a frequency effect, not a thermal effect, so Answers C and D are incorrect. **Answer A** gives the correct definition, so choose this option. Answer B has the opposite relationship with frequency, so it is incorrect.

E5D02 Why is it important to keep lead lengths short for components used in circuits for VHF and above?
  A. To increase the thermal time constant
  B. To avoid unwanted inductive reactance
  C. To maintain component lifetime
  D. All of these choices are correct

Electronic leads can act as antennas and have inductive effects. The unwanted inductive reactance in **Answer B** is the right choice. Answers A and C are not electrically relevant, so they are incorrect, as well as Answer D.

E5D03 What is microstrip?
  A. Lightweight transmission line made of common zip cord
  B. Miniature coax used for low power applications
  C. Short lengths of coax mounted on printed circuit boards to minimize time delay between microwave circuits
  D. Precision printed circuit conductors above a ground plane that provide constant impedance interconnects at microwave frequencies

Microstrips are a form of printed circuit board with a ground plane for microwave applications as in **Answer D**. The other choices are distractions.

E5D04 Why are short connections necessary at microwave frequencies?
- A. To increase neutralizing resistance
- B. To reduce phase shift along the connection
- C. To increase compensating capacitance
- D. To reduce noise figure

This question is an extension of the previous question. Because connections can have inductive characteristics that increase with length, and the reactance scales with frequency, keeping the connections short at microwave frequencies is a good recommendation. Having the inductive components will induce a phase shift, so **Answer B** is the right choice.

E5D05 What is the power factor of an RL circuit having a 30 degree phase angle between the voltage and the current?
- A. 1.73
- B. 0.5
- C. 0.866
- D. 0.577

The power factor is the cosine of the phase angle: $PF = \cos 30° = 0.866$. This result makes **Answer C** the correct choice. Other than Answer A, which is impossible, the other choices correspond to other angles.

E5D06 In what direction is the magnetic field oriented about a conductor in relation to the direction of electron flow?
- A. In the same direction as the current
- B. In a direction opposite to the current
- C. In all directions; omni-directional
- D. In a circle around the conductor

Mathematically, we define the direction using a cross-product relationship, making for a circular flow. In this case, **Answer D** describes the flow direction, so this is the correct choice for this question. Answers A, B, and C are electrically incorrect.

E5D07 How many watts are consumed in a circuit having a power factor of 0.71 if the apparent power is 500VA?
- A. 704 W
- B. 355 W
- C. 252 W
- D. 1.42 mW

Here we apply the PF to the power computation. The power is $P = (500)(0.71) = 355$ W as in **Answer B**. The other choices are to distract you.

E5D08 How many watts are consumed in a circuit having a power factor of 0.6 if the input is 200VAC at 5 amperes?
  A. 200 watts
  B. 1000 watts
  C. 1600 watts
  D. 600 watts

The power is $P = V \times I \times PF = (200)(5)(0.6) = 600\,\text{W}$ as in **Answer D**. Answer B is the power without the power factor applied, so it is incorrect. The other choices have math mistakes.

E5D09 What happens to reactive power in an AC circuit that has both ideal inductors and ideal capacitors?
  A. It is dissipated as heat in the circuit
  B. It is repeatedly exchanged between the associated magnetic and electric fields, but is not dissipated
  C. It is dissipated as kinetic energy in the circuit
  D. It is dissipated in the formation of inductive and capacitive fields

The reactive power remains in the electric and magnetic fields, as in **Answer B**, so that is the correct choice. The power dissipated as heat is the characteristic of a non-ideal, resistive device, so Answer A cannot be correct. Electrical circuits do not have kinetic energy unless you throw them, so Answer C cannot be correct either. Answer D is incorrect because the establishment of the field is part of the initial transient response of the circuit.

E5D10 How can the true power be determined in an AC circuit where the voltage and current are out of phase?
  A. By multiplying the apparent power times the power factor
  B. By dividing the reactive power by the power factor
  C. By dividing the apparent power by the power factor
  D. By multiplying the reactive power times the power factor

The correct procedure is to multiply the apparent power by the power factor, as given in **Answer A**. Answer D uses the wrong power measurement, so it is incorrect. Answers B and C are not the correct procedure for using the power factor, so they are wrong.

E5D11 What is the power factor of an RL circuit having a 60 degree phase angle between the voltage and the current?
  A. 1.414
  B. 0.866
  C. 0.5
  D. 1.73

The power factor is $PF = \cos 60° = 0.5$, which makes **Answer C** the correct choice. Answers A and D cannot be valid power factors because they are greater than one. Answer B is for a $30°$ phase angle.

E5D12 How many watts are consumed in a circuit having a power factor of 0.2 if the input is 100VAC at 4 amperes?
  A. 400 watts
  B. 80 watts
  C. 2000 watts
  D. 50 watts

Again, we apply the power factor to compute the power from $P = V \times I \times PF = (100)(4)(0.2) = 80\,\text{W}$. **Answer B** is the correct choice for this question. Answer A ignores the power factor, so it is not a correct choice. The other choices contain math mistakes

E5D13 How many watts are consumed in a circuit consisting of a 100-ohm resistor in series with a 100-ohm inductive reactance drawing 1 ampere?
  A. 70.7 Watts
  B. 100 Watts
  C. 141.4 Watts
  D. 200 Watts

First, we need to find the phase angle between the voltage and the current. The total impedance is $Z = 100\,\Omega + j100\,\Omega = 141.42\,\Omega$ at an angle of $45°$. The power factor is $PF = \cos 45° = 0.707$. The power is $P = I^2|Z|PF = (1)(141.42)(0.707) = 100\,\text{W}$. **Answer B** is the correct answer.

E5D14 What is reactive power?
  A. Wattless, nonproductive power
  B. Power consumed in wire resistance in an inductor
  C. Power lost because of capacitor leakage
  D. Power consumed in circuit Q

The apparent power in a reactive load is the reactive power, which is wattless, nonproductive power, so **Answer A** is the correct choice. Do not be confused with Answer B since it concerns a reactive element, but this is not the definition of reactive power. Answer C is a concern in a circuit, but this is not the definition of reactive power. Answer D is a silly distraction.

E5D15 What is the power factor of an RL circuit having a 45-degree phase angle between the voltage and the current?
    A. 0.866
    B. 1.0
    C. 0.5
    D. 0.707

The power factor is $\cos\theta$. In this case, $PF = \cos 45° = 0.707$ making **Answer D** the correct choice. The other choices correspond to angles other than $45°$, so they are incorrect.

Chapter 6

E6 — CIRCUIT COMPONENTS

6.1 Introduction

The *Circuit Components* subelement covers many more details of circuit components than we saw in the Technician and General Class examination questions. The questions will cover semiconductors extensively and discrete analog, digital, and optical devices. The *Circuit Components* subelement has the following question groups:

A. Semiconductor materials and devices
B. Diodes
C. Digital ICs
D. Toroidal and solenoidal inductors
E. Analog ICs
F. Electro-optical technology

Subelement 6 will generate six questions on the Extra examination.

6.2 Radio Engineering Concepts

Diodes and Transistors The Extra Class examination builds on the knowledge needed for the General Class examination. Both diodes and transistors use the characteristics of the semiconductor materials to achieve their functions. Engineers classify semiconductor material as "N-type" if it has a surplus of free electrons and "P-type" if it has a deficit of electrons, also known as having holes. As a review, diodes are devices to allow current flow in a single direction. When they are "forward biased," they permit current flow, and when "reverse biased," they prevent current flow. In the earlier study guides, we looked at semiconductors based on Silicon (Si) and Germanium (Ge). Other more

Table 6.1: Diode types.

Type	Notes
Si	silicon diode with a forward voltage drop of 0.7 V
Ge	germanium diode with a forward voltage drop of 0.3 V
Hot Carrier	a Schottky diode
LED	emits light when forward biased
PIN	a P-type – Intrinsic Material – N-type junction configuration used in switching
Point Contact	diode with an embedded wire to improve switching time
Schottky	rapid switching with a forward voltage drop of 0.2 V
SCR	Silicon Controlled Rectifier; a type of thyristor
Thyristor	uses a small gate current to open the flow of current through the diode
Triac	a form of thyristor diode
Tunnel	exhibits a negative resistance (current decreases as voltage increases) region; used for high-speed switching
Varactor	diode with a large junction capacitance range that can be controlled
Zener	permits current flow at a constant voltage when reverse biased

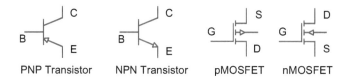

Figure 6.1: BJT and MOSFET transistor types and the associated schematic symbols with the connector labels.

modern technologies, such as Gallium Arsenide (GaAs), have better properties, especially for high-speed applications. Table 6.1 lists semiconductor diodes types that you need to know about for the Extra Class examination.

As we saw in the General Class study guide, two transistor classes appear in the license examination questions: the Bipolar Junction Transistor (BJT) and the Metal Oxide Semiconductor Field Effect Transistor (MOSFET). The type of semiconductor material the designer uses in fabricating the transistor controls the transistor's exact characteristics. As we mentioned above, designers use two semiconductor types: N-type, which has excess electrons for conducting electricity, and P-type, which has an excess of positive material. Figure 6.1 illustrates the schematic symbols for the BJT and MOSFET transistors. The connectors for the BJT are Emitter (E), Collector (C), and Base (B), while the connectors for the MOSFET are Source (S), Drain (D), and Gate (G). For the

BJT, the base-to-emitter junction acts as a diode, so when the circuit turns on the transistor, the voltage between them is similar, for example, 0.7 V for a Si transistor.

Engineers define the BJT current gains in terms of the changes (Δ) in the collector current, I_C, the base current, I_B, and emitter current, I_E, as

$$\alpha = -\frac{\Delta I_C}{\Delta I_E}$$

$$\beta = \frac{\Delta I_C}{\Delta I_B}$$

$$\alpha = -\frac{\beta}{1+\beta}$$

There are questions in the examination pool on these current gains.

Digital Devices In addition to the basic logic gates we saw in the General Class study guide, there are other digital circuits we need to know. The first of these is the comparator. Figure 6.2 shows a typical circuit configuration. The device changes its output state based on the input voltage difference. The op-amp's positive input is from a voltage divider that sets the reference voltage. The op-amp's negative input is the variable input voltage. The output swings between the high and low supply voltages based on the voltage difference on the op-amp's terminals. The compara-

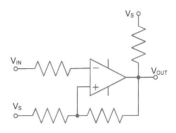

Figure 6.2: Typical comparator circuit configuration.

tor usually has internal hysteresis circuitry to prevent small noise fluctuations from causing rapid changes in the output. The output has a "pull-up" resistor to aid interfacing with other circuits. The comparator functions as a 1-bit Analog-to-Digital Converter (ADC).

When we saw logic devices earlier, we showed a truth table with outputs taking on one of two states: logic 0 or a logic 1. There is a third output type called "tri-state" logic, where the output is not a logic state but very high resistance until the logic output is enabled. This property permits multiple outputs to share a common output line, and then the user can select one as needed.

Logic gates form a low-level building block. In modern circuitry, one does not always design with individual gates. Designers also use higher integration with packages such as a

Programmable Logic Device — the generic name for a digital device that the user can program to represent logic circuits and functions

Programmable Gate Array — a type of programmable logic device composed of multiple logic blocks and the interconnections; also called a Field Programmable Gate Array

Inductor Design   The examination has several questions on designing inductors with toroidal cores. The design equation is in terms of the numbers of turns, N, and the permeability, A_L, of the core material. The designer computes the inductance, L, using $L = N^2 A_L$.

Optoisolators and Shaft Encoders   Optoisolators and optocouplers are photonic devices designers use to provide electrical isolation between circuit domains. The designer can build them with a Light Emitting Diode (LED) on the input side to generate the signal. A phototransistor on the output side receives the pulses and transfers them to the second circuit, as Figure 6.3 shows. The manufacturer can mount the LED and phototransistor in the same package. The devices keep noise from going from one circuit to the other. Designers can use them to send signals between circuits on separate power and ground paths.

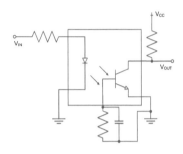

Figure 6.3: Typical optoisolator circuit configuration.

If you can imagine breaking into the optoisolator configuration and placing a temporary block to break the photon flow from the LED to the phototransistor, then you can imagine a new device. This temporary blocking is the principle behind a shaft encoder. The shaft has a wheel with a coded pattern of transparent and opaque areas that match the rotational position of the shaft. As the shaft turns, a detector can read the pattern to determine the shaft's position and direction of motion.

6.3   E6A - Semiconductor Materials and Devices

6.3.1   Overview

The *Semiconductor Materials and Devices* question group in Subelement E6 tests you on aspects of semiconductor configuration and characteristics. The questions will also cover transistors and other semiconductor devices' properties and circuit schematic symbols. The *Semiconductor Materials and Devices* group covers topics such as

- Semiconductor materials: germanium, silicon, P-type, N-type
- Transistor types: NPN, PNP, junction, and field-effect transistors
- Enhancement mode and depletion mode
- MOS and CMOS
- N-channel and P-channel

The test producer will select one of the 12 questions in this group for your exam.

6.3.2 Questions

E6A01 In what application is gallium arsenide used as a semiconductor material?
  A. In high-current rectifier circuits
  B. In high-power audio circuits
  C. In microwave circuits
  D. In very low-frequency RF circuits

Since we deal with microwave-frequency communications, that is a hint that **Answer C** is correct. In Answers A, B, or D, designers prefer germanium or silicon to gallium arsenide for those applications, so they are not good choices.

E6A02 Which of the following semiconductor materials contains excess free electrons?
  A. N-type
  B. P-type
  C. Bipolar
  D. Insulated gate

From your science classes, you will remember that electrons carry a negative charge or "N." If we have excess electrons, we have N-type material making **Answer A** the correct choice. Having more positive charges makes the material P-type. The bipolar and insulated gate options in Answers C and D do not tell you what type of materials are in the semiconductor, so they are not good choices.

E6A03 Why does a PN-junction diode not conduct current when reverse biased?
  A. Only P-type semiconductor material can conduct current
  B. Only N-type semiconductor material can conduct current
  C. Holes in P-type material and electrons in the N-type material are separated by the applied voltage, widening the depletion region
  D. Excess holes in P-type material combine with the electrons in N-type material, converting the entire diode into an insulator

When the diode is reverse biased, the applied voltage separates the electrons and the holes so current does not flow, which makes **Answer C** the correct choice. Answers A, B, and D are incorrect statements.

E6A04 What is the name given to an impurity atom that adds holes to a semiconductor crystal structure?
  A. Insulator impurity
  B. N-type impurity
  C. Acceptor impurity
  D. Donor impurity

Because holes can gobble up free electrons, they are known as "acceptors," which makes **Answer C** the correct choice. N-type material adds electrons and not holes. The term "donors" alone does not specify if the designer adds electrons or holes. Answer A is to distract you.

E6A05 How does DC input impedance at the gate of a field-effect transistor compare with the DC input impedance of a bipolar transistor?
   A. They are both low impedance
   B. An FET has lower input impedance
   C. An FET has higher input impedance
   D. They are both high impedance

The transistor impedance is a function of the materials and techniques used in their construction. The Field Effect Transistor (FET) and BJT have different impedances. The FET is high impedance, while the BJT is low impedance making **Answer C** the correct choice. Answer B is the reverse, so be careful when reading.

E6A06 What is the beta of a bipolar junction transistor?
   A. The frequency at which the current gain is reduced to 0.707
   B. The change in collector current with respect to base current
   C. The breakdown voltage of the base to collector junction
   D. The switching speed

This question asks about another term you may need to memorize if electronics is not your specialty. The beta (β) is the change in collector current with respect to the base current, as in **Answer B**.

E6A07 Which of the following indicates that a silicon NPN junction transistor is biased on?
   A. Base-to-emitter resistance of approximately 6 to 7 ohms
   B. Base-to-emitter resistance of approximately 0.6 to 0.7 ohms
   C. Base-to-emitter voltage of approximately 6 to 7 volts
   D. Base-to-emitter voltage of approximately 0.6 to 0.7 volts

When the circuit biases a silicon NPN transistor on, the base-to-emitter voltage is typically 0.6 to 0.7 V, as in **Answer D**. Be careful with Answer B because it has the same numbers but the wrong units. The other choices are distractions.

E6A08 What term indicates the frequency at which the grounded-base current gain of a transistor has decreased to 0.7 of the gain obtainable at 1 kHz?
   A. Corner frequency
   B. Alpha rejection frequency
   C. Beta cutoff frequency
   D. Alpha cutoff frequency

The device is typically in the cutoff region when looking at a decreasing gain, so Answers C and D are candidates for the correct choice. This specific definition is for the alpha cutoff frequency, so **Answer D** is the right choice.

E6A09 What is a depletion-mode FET?
   A. An FET that exhibits a current flow between source and drain when no gate voltage is applied
   B. An FET that has no current flow between source and drain when no gate voltage is applied
   C. Any FET without a channel
   D. Any FET for which holes are the majority carriers

A depletion-mode FET has a current flow between the source and the drain even when the circuit does not apply a gate voltage, as in **Answer A**. The other choices are to distract you from this definition.

Figure E6-1

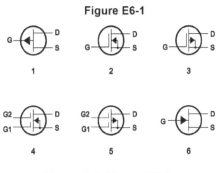

Figure 6.4: Figure E6-1.

E6A10 In Figure E6-1, what is the schematic symbol for an N-channel dual-gate MOSFET?
   A. 2
   B. 4
   C. 5
   D. 6

Symbol 1 is a P-channel Junction FET, while Symbol 2 is a single-gate N-channel MOSFET. Symbol 3 is a single-gate P-channel MOSFET, and Symbol 4 is a dual-gate N-channel MOSFET. Symbol 5 is a dual-gate P-channel MOSFET, while Symbol 6 is an N-channel Junction FET. Therefore, the correct choice is **Answer B**. The others are to distract you.

E6A11 In Figure E6-1, what is the schematic symbol for a P-channel junction FET?
   A. 1
   B. 2
   C. 3
   D. 6

Symbol 1 is a P-channel Junction FET, while Symbol 2 is a single-gate N-channel MOSFET. Symbol 3 is a single-gate P-channel MOSFET, and Symbol 4 is a dual-gate N-channel MOSFET. Symbol 5 is a dual-gate P-channel MOSFET, while Symbol 6 is an N-channel Junction FET. Therefore, the correct choice is **Answer A**. The others are to distract you.

E6A12 Why do many MOSFET devices have internally connected Zener diodes on the gates?
- A. To provide a voltage reference for the correct amount of reverse-bias gate voltage
- B. To protect the substrate from excessive voltages
- C. To keep the gate voltage within specifications and prevent the device from overheating
- D. To reduce the chance of static damage to the gate

Answer B is close to the right answer, but it is not the best choice. The prevention of static damage, as given in **Answer D**, is the best choice to answer this question. Answers A and C are to distract you.

6.4   E6B - Diodes

6.4.1   Overview

The *Diodes* question group in Subelement E6 tests on the operating characteristics and schematic symbols for several families of diodes found in radio circuits. The test producer will select one of the 11 questions in this group for your exam.

6.4.2   Questions

E6B01 What is the most useful characteristic of a Zener diode?
- A. A constant current drop under conditions of varying voltage
- B. A constant voltage drop under conditions of varying current
- C. A negative resistance region
- D. An internal capacitance that varies with the applied voltage

Engineers sometimes call Zener diodes a "poor man's voltage regulator" because of their ability to maintain a constant voltage, so **Answer B** is the right choice. Answer A has the voltage and current interchanged, so it is incorrect. Answer C is for a tunnel diode, which is incorrect as well. Answer D is not a characteristic of the Zener diode but the varactor diode.

E6B02 What is an important characteristic of a Schottky diode as compared to an ordinary silicon diode when used as a power supply rectifier?
- A. Much higher reverse voltage breakdown
- B. More constant reverse avalanche voltage
- C. Longer carrier retention time
- D. Less forward voltage drop

The characteristic to remember is that the Schottky diode has less forward voltage drop than an ordinary diode. This fact makes **Answer D** the right choice.

E6B03 What type of bias is required for an LED to emit light?
  A. Reverse bias
  B. Forward bias
  C. Zero bias
  D. Inductive bias

The LED "glows" when the junction is forward biased, so **Answer B** is the right choice. A LED is a junction diode, so it does not emit light when the circuit reverse biases the junction making Answer A incorrect. Answers C and D are technobabble to try to confuse you.

E6B04 What type of semiconductor device is designed for use as a voltage-controlled capacitor?
  A. Varactor diode
  B. Tunnel diode
  C. Silicon-controlled rectifier
  D. Zener diode

This question has the definition of a varactor diode, so **Answer A** is the right choice. We already saw the characteristics of Zener and tunnel diodes, so we should be able to spot them as distractions for this question.

E6B05 What characteristic of a PIN diode makes it useful as an RF switch?
  A. Extremely high reverse breakdown voltage
  B. Ability to dissipate large amounts of power
  C. Reverse bias controls its forward voltage drop
  D. Low junction capacitance

The large Intrinsic (I) region between the P-type and N-type material gives the Positive-Intrinsic-Negative (PIN) diode low junction capacitance. **Answer D** is the right choice. The others are not properties of a PIN diode.

E6B06 (D) Which of the following is a common use of a Schottky diode?
  A. As a rectifier in high current power supplies description
  B. As a variable capacitance in an automatic frequency control circuit
  C. As a constant voltage reference in a power supply
  D. As a VHF/UHF mixer or detector

Schottky diodes have their primary application in high-speed circuitry. A Very High Frequency (VHF) or Ultra High Frequency (UHF) mixer requires high-speed circuit elements, so **Answer D** is the appropriate choice. The other options do not typically require a high-speed diode in their circuitry.

E6B07 What is the failure mechanism when a junction diode fails due to excessive current?
  A. Excessive inverse voltage
  B. Excessive junction temperature
  C. Insufficient forward voltage
  D. Charge carrier depletion

As more current flows, the junction will warm up, so **Answer B** is the correct reason. Excessive current will not produce an inverse voltage or require an excessive forward bias voltage. The charge carrier depletion is a regular property of operating diodes and does not lead to failure.

E6B08 Which of the following is a Schottky barrier diode?
  A. Metal-semiconductor junction
  B. Electrolytic rectifier
  C. PIN junction
  D. Thermionic emission diode

Answer A provides the correct response: the metal-semiconductor junction. The other options are to distract you.

E6B09 What is a common use for point-contact diodes?
  A. As a constant current source
  B. As a constant voltage source
  C. As an RF detector
  D. As a high-voltage rectifier

Designers use point contact diodes in Radio Frequency (RF) detection circuits, so **Answer C** is correct. Designers use other circuits for the applications in Answers A and B. Designers often use silicon diodes for Answer D.

E6B10 In Figure E6-2, what is the schematic symbol for a light-emitting diode?
  A. 1
  B. 5
  C. 6
  D. 7

Figure E6-2

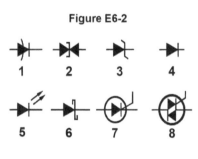

Figure 6.5: Figure E6-2.

The LED emits light, so use the arrows coming from the diode to key into Symbol 5 in **Answer B** as indicating the right choice. Symbol 1 is a voltage variable capacitor; Symbol 2 is a tunnel diode; Symbol 3 is a Zener diode; Symbol 4 is a diode; Symbol 6 is a

Schottky diode; Symbol 7 is a thyristor, and Symbol 8 is a triac.

E6B11 What is used to control the attenuation of RF signals by a PIN diode?
  A. Forward DC bias current
  B. A sub-harmonic pump signal
  C. Reverse voltage larger than the RF signal
  D. Capacitance of an RF coupling capacitor

The circuit uses a forward Direct Current (DC) bias current for the control, as in **Answer A**, so that is correct.

6.5 E6C - Digital ICs

6.5.1 Overview

The *Digital ICs* question group in Subelement E6 covers digital devices, including logic gates, schematic symbols, and interfacing options such as tri-state logic. The *Digital ICs* group covers topics such as
  • Families of digital ICs
  • Gates
  • Programmable Logic Devices (PLDs)
The test producer will select one of the 11 questions in this group for your exam.

6.5.2 Questions

E6C01 What is the function of hysteresis in a comparator?
  A. To prevent input noise from causing unstable output signals
  B. To allow the comparator to be used with AC input signal
  C. To cause the output to change states continually
  D. To increase the sensitivity

A comparator, such as the one in Figure 6.2, looks at the difference in the inputs for determining the output. This type of differential examination is susceptible to noise. Hysteresis provides voltage levels that the device must reach before changing state so that noise is not a strong influence. This property makes **Answer A** the correct choice, and the other options are untrue statements.

E6C02 What happens when the level of a comparator's input signal crosses the threshold?
  A. The IC input can be damaged
  B. The comparator changes its output state
  C. The comparator enters latch-up
  D. The feedback loop becomes unstable

Because we are talking about digital devices, the inputs and outputs have defined states. When the input voltage crosses the threshold, the output changes
state, as in **Answer B**. The other options are untrue statements.

E6C03 What is tri-state logic?
A. Logic devices with 0, 1, and high-impedance output states
B. Logic devices that utilize ternary math
C. Low-power logic devices designed to operate at 3 volts
D. Proprietary logic devices manufactured by Tri-State Devices

So far, we have only considered the standard bi-state logic of a "0" and "1."
However, there exists a very useful logic family called tri-state, where the third
state is a high-impedance output to permit innovative connections and inter-
faces. **Answer A** captures this definition. The other options are to distract you.

E6C04 Which of the following is an advantage of BiCMOS logic?
A. Its simplicity results in much less expensive devices than standard CMOS
B. It is totally immune to electrostatic damage
C. It has the high input impedance of CMOS and the low output impedance
of bipolar transistors
D. All of these choices are correct

Answers A and B are untrue statements, making Answer D incorrect. BiCMOS
combines the high input impedance of Complementary Metal Oxide Semiconductor (CMOS) with the low output impedance of BJTs, as in **Answer C**.

E6C05 What is an advantage of CMOS logic devices over TTL devices?
A. Differential output capability
B. Lower distortion
C. Immune to damage from static discharge
D. Lower power consumption

The big advantage of CMOS over Transistor-Transistor Logic (TTL), especially
with modern circuitry going to lower power supply voltages, is the lower power
consumption of CMOS. **Answer D** is the right choice.

E6C06 Why do CMOS digital integrated circuits have high immunity to noise on
the input signal or power supply?
A. Large bypass capacitance is inherent
B. The input switching threshold is about two times the power supply voltage
C. The input switching threshold is about one-half the power supply voltage
D. Bandwidth is very limited

The high switching threshold given in **Answer C** provides input noise immunity,
so this is the right choice. The other choices are not electrically correct.

E6C07 What best describes a pull-up or pull-down resistor?
A. A resistor in a keying circuit used to reduce key clicks
B. A resistor connected to the positive or negative supply line used to establish a voltage when an input or output is an open circuit
C. A resistor that ensures that an oscillator frequency does not drift
D. A resistor connected to an op-amp output that prevents signals from exceeding the power supply voltage

Designers connect these resistors to either the positive or negative supply voltage to set an input or output voltage when that terminal is open circuit. This use makes **Answer B** the correct choice, while the other statements are wrong.

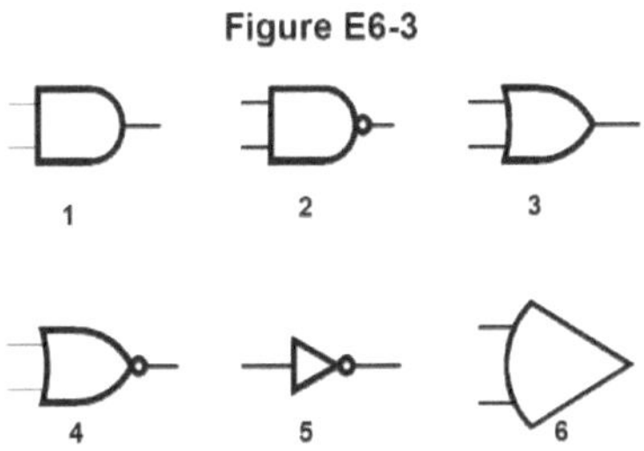

Figure E6-3

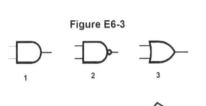

Figure 6.6: Figure E6-3.

E6C08 In Figure E6-3, what is the schematic symbol for a NAND gate?
A. 1
B. 2
C. 3
D. 4

Symbol 1 is an AND gate, Symbol 2 is a NAND gate, Symbol 3 is an OR gate, Symbol 4 is a NOR gate, Symbol 5 is a NOT gate, and Symbol 6 is a buffer. Therefore, the correct choice is **Answer B**.

E6C09 What is a Programmable Logic Device (PLD)?
A. A logic circuit that can be modified during use
B. A programmable collection of logic gates and circuits in a single integrated circuit
C. Programmable equipment used for testing digital logic integrated circuits
D. A type of transistor whose gain can be changed by digital logic circuits

A Programmable Logic Device (PLD) is a single integrated set of programmable logic gates and circuits, as in **Answer B**. The other statements are to distract you.

E6C10 In Figure E6-3, what is the schematic symbol for a NOR gate?
A. 1
B. 2
C. 3
D. 4

Symbol 1 is an AND gate, Symbol 2 is a NAND gate, Symbol 3 is an OR gate, Symbol 4 is a NOR gate, Symbol 5 is a NOT gate, and Symbol 6 is a buffer. Therefore, the correct choice is **Answer D**.

E6C11 In Figure E6-3, what is the schematic symbol for the NOT operation (inverter)?
  A. 2
  B. 4
  C. 5
  D. 6

Symbol 1 is an AND gate, Symbol 2 is a NAND gate, Symbol 3 is an OR gate, Symbol 4 is a NOR gate, Symbol 5 is a NOT gate, also known as an inverter, and Symbol 6 is a buffer. Therefore, the correct choice is **Answer C**.

6.6   E6D - Toroidal and Solenoidal Inductors

6.6.1   Overview

The *Toroidal and Solenoidal Inductors* question group in Subelement E6 covers inductors, including those with different core configurations, quartz crystals, and their use in circuits. The *Toroidal and Solenoidal Inductors* group covers topics such as
  • Permeability, core material, selecting, and winding
  • Transformers
  • Piezoelectric devices
The test producer will select one of the 13 questions in this group for your exam.

6.6.2   Questions

E6D01 Why should core saturation of an impedance matching transformer be avoided?
  A. Harmonics and distortion could result
  B. Magnetic flux would increase with frequency
  C. RF susceptance would increase
  D. Temporary changes of the core permeability could result

If the core is saturated, harmonics and other bad effects can occur, which makes **Answer A** the correct choice. The other choices are incorrect statements.

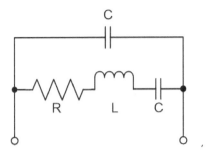

Figure 6.7: Equivalent circuit model for a crystal.

E6D02 What is the equivalent circuit of a quartz crystal?
  A. Motional capacitance, motional inductance, and loss resistance in series, all in parallel with a shunt capacitor representing electrode and stray capacitance
  B. Motional capacitance, motional inductance, loss resistance, and a capacitor representing electrode and stray capacitance all in parallel
  C. Motional capacitance, motional inductance, loss resistance, and a capacitor representing electrode and stray capacitance all in series
  D. Motional inductance and loss resistance in series, paralleled with motional capacitance and a capacitor representing electrode and stray capacitance

Figure 6.7 illustrates the circuit model for a crystal which corresponds to **Answer A**. Be careful because all choices look similar.

E6D03 Which of the following is an aspect of the piezoelectric effect?
  A. Mechanical deformation of material by the application of a voltage
  B. Mechanical deformation of material by the application of a magnetic field
  C. Generation of electrical energy in the presence of light
  D. Increased conductivity in the presence of light

The piezoelectric effect is related to the mechanical deformation of the crystal and the voltage causing it, so **Answer A** is the correct choice. Answer B is not this physical effect, so this is an incorrect choice. Answer C is a form of the photoelectric effect, so this is wrong. Answer D is photoconductivity.

E6D04 Which materials are commonly used as a core in an inductor?
  A. Polystyrene and polyethylene
  B. Ferrite and brass
  C. Teflon and Delrin
  D. Cobalt and aluminum

Answers A, C, and D do not contain materials with favorable magnetic properties, so they are incorrect. Ferrite and brass are the typical choices, so **Answer B** is correct. Manufacturers sometimes use aluminum in high-frequency applications, but it is not as common as ferrite.

E6D05 What is one reason for using ferrite cores rather than powdered iron in an inductor?
  A. Ferrite toroids generally have lower initial permeability
  B. Ferrite toroids generally have better temperature stability
  C. Ferrite toroids generally require fewer turns to produce a given inductance value
  D. Ferrite toroids are easier to use with surface mount technology

Based on the design equation $L = N^2 A_L$ and the relative permeabilities, A_L, we should be able to spot **Answer C** as the correct one. Answer A is wrong because ferrite toroids have higher permeabilities. Answer B is incorrect because powdered iron has better temperature stability.

E6D06 What core material property determines the inductance of an inductor?
  A. Thermal impedance
  B. Resistance
  C. Reactivity
  D. Permeability

The permeability of the material, A_L, determines the inductance making **Answer D** the correct choice. Resistance is not involved. Reactance is based on the inductance and not the other way around.

E6D07 What is current in the primary winding of a transformer called if no load is attached to the secondary?
  A. Magnetizing current
  B. Direct current
  C. Excitation current
  D. Stabilizing current

This current establishes the magnetic field, and engineers call it the "magnetizing current," as in **Answer A.** The other choices are not correct in this situation.

E6D08 What is one reason for using powdered-iron cores rather than ferrite cores in an inductor?
  A. Powdered-iron cores generally have greater initial permeability
  B. Powdered-iron cores generally maintain their characteristics at higher currents
  C. Powdered-iron cores generally require fewer turns to produce a given inductance
  D. Powdered-iron cores use smaller diameter wire for the same inductance

Because an inductor can become warm when current flows, temperature stability is crucial, which makes **Answer B** the best choice among those given. The other options represent incorrect statements.

E6D09 What devices are commonly used as VHF and UHF parasitic suppressors at the input and output terminals of a transistor HF amplifier?
A. Electrolytic capacitors
B. Butterworth filters
C. Ferrite beads
D. Steel-core toroids

The correct answer is ferrite beads, so choose **Answer C**. Electrolytic capacitors are not parasitic suppressors. Designers make Butterworth filters with Resistor-Inductor-Capacitor (RLC) components, and they are not parasitic suppressors. Generally, manufacturers do not use steel as a toroidal material.

E6D10 What is a primary advantage of using a toroidal core instead of a solenoidal core in an inductor?
A. Toroidal cores confine most of the magnetic field within the core material
B. Toroidal cores make it easier to couple the magnetic energy into other components
C. Toroidal cores exhibit greater hysteresis
D. Toroidal cores have lower Q characteristics

The reason for using the toroidal core is to confine the magnetic field, so **Answer A** is the right choice. Magnetic coupling is the opposite of the desired effect. Hysteresis is generally not a suitable property for this application. Answer D is irrelevant to the purpose of confining the magnetic field.

E6D11 Which type of core material decreases inductance when inserted into a coil?
A. Ceramic
B. Brass
C. Ferrite
D. Powdered iron

Of the choices given, designers use the brass of **Answer B** to reduce inductance, so that is the right choice.

E6D12 What is inductor saturation?
A. The inductor windings are over-coupled
B. The inductor's voltage rating is exceeded causing a flashover
C. The ability of the inductor's core to store magnetic energy has been exceeded
D. Adjacent inductors become over-coupled

The magnetic field in the inductor can only hold so much energy. When that field is "filled" to its maximum capacity, it is "saturated," as in **Answer C**. The other choices are distractions.

E6D13 What is the primary cause of inductor self-resonance?
A. Inter-turn capacitance
B. The skin effect
C. Inductive kickback
D. Non-linear core hysteresis

We need both an inductor and a capacitor to make a resonant circuit. Look for the inter-turn capacitance in **Answer A** as the clue to the right choice.

6.7   E6E - Analog ICs

6.7.1   Overview

The *Analog ICs* question group in Subelement E6 looks at the operational characteristics of multiple analog devices in radio circuit applications. The *Analog ICs* group covers topics such as MMICs and IC packaging characteristics. The test producer will select one of the 12 questions in this group for your exam.

6.7.2   Questions

E6E01 Why is gallium arsenide (GaAs) useful for semiconductor devices operating at UHF and higher frequencies?
A. Higher noise figures
B. Higher electron mobility
C. Lower junction voltage drop
D. Lower transconductance

Designers often use Gallium Arsenide (GaAs) components in UHF and higher-frequency applications. The higher electron mobility of **Answer B** enables this utility. The other properties are not specific to high-frequency applications.

E6E02 Which of the following device packages is a through-hole type?
A. DIP
B. PLCC
C. Ball grid array
D. SOT

The Dual In-line Package (DIP) format uses holes drilled through the circuit board to make the attachment, so **Answer A** is the correct choice. Ball grid array and Plastic-Leaded Chip Carrier (PLCC) are chip carriers on the circuit board's surface, and Small Outline Transistor (SOT) is a surface-mounted transistor.

E6E03 Which of the following materials is likely to provide the highest frequency of operation when used in MMICs?
  A. Silicon
  B. Silicon nitride
  C. Silicon dioxide
  D. Gallium nitride

Silicon is not the material of choice for modern high-frequency circuits. Silicon dioxide is an insulator and not a circuit conductor. Gallium Nitride (GaN) is a high-frequency material, so **Answer D** is correct. Be careful with Answer B because it looks close, but SiN is a ceramic material.

E6E04 Which is the most common input and output impedance of circuits that use MMICs?
  A. 50 ohms
  B. 300 ohms
  C. 450 ohms
  D. 10 ohms

Because Monolithic Microwave Integrated Circuits (MMICs) frequently interface with RF devices, you may suspect that the correct answer is 50 Ω, as in **Answer A**. That is the right choice, while the other choices are to distract you.

E6E05 Which of the following noise figure values is typical of a low-noise UHF preamplifier?
  A. 2 dB
  B. -10 dB
  C. 44 dBm
  D. -20 dBm

Engineers rate noise figures in "dB" units and not "dBm" units, so Answers C and D are incorrect. A typical "low-noise" noise figure value for this type of amplifier is around 2 dB, which makes **Answer A** the correct choice.

E6E06 What characteristics of the MMIC make it a popular choice for VHF through microwave circuits?
  A. The ability to retrieve information from a single signal even in the presence of other strong signals
  B. Plate current that is controlled by a control grid
  C. Nearly infinite gain, very high input impedance, and very low output impedance
  D. Controlled gain, low noise figure, and constant input and output impedance over the specified frequency range

MMIC packaging permits designers to have good gain control, low noise, and constant impedances, as in **Answer D**. The other choices are incorrect.

E6E07 What type of transmission line is used for connections to MMICs?
  A. Miniature coax
  B. Circular waveguide
  C. Parallel wire
  D. Microstrip

MMICs utilize integrated circuit construction techniques. A microstrip is a form of integrated circuit construction, so **Answer D** is the right choice. Designers use the other choices in different applications.

E6E08 How is power supplied to the most common type of MMIC?
  A. Through a resistor and/or RF choke connected to the amplifier output lead
  B. MMICs require no operating bias
  C. Through a capacitor and RF choke connected to the amplifier input lead
  D. Directly to the bias voltage (VCC IN) lead

The correct method for the power connection is through a resistor or a RF choke, as described in **Answer A**. The others are distractions.

E6E09 Which of the following component package types would be most suitable for use at frequencies above the HF range?
  A. TO-220
  B. Axial lead
  C. Radial lead
  D. Surface mount

Above the High Frequency (HF) range, modern designers tend to use surface mount techniques, so **Answer D** is the right choice. The other choices have relatively large leads that are not optimal for the higher frequency signals.

E6E10 What advantage does surface-mount technology offer at RF compared to using through-hole components?
A. Smaller circuit area
B. Shorter circuit-board traces
C. Components have less parasitic inductance and capacitance
D. All these choices are correct

Each statement in Answers A, B, and C is a characteristic of surface-mount technology, which makes **Answer D** the best choice.

E6E11 What is a characteristic of DIP packaging used for integrated circuits?
A. Package mounts in a direct inverted position
B. Low leakage doubly insulated package
C. Two chips in each package (Dual In Package)
D. A total of two rows of connecting pins placed on opposite sides of the package (Dual In-line Package)

The Dual In-line Package (DIP) name tells the configuration: two rows of pins in parallel on either side of the package. **Answer D** is the right choice, while the other options are distractions.

E6E12 Why are DIP through-hole package ICs not typically used at UHF and higher frequencies?
A. Too many pins
B. Epoxy coating is conductive above 300 MHz
C. Excessive lead length
D. Unsuitable for combining analog and digital signals

At UHF and higher frequencies, short lead length is a "must" for effective operation. **Answer C** is the right choice, while the others are not correct statements for this application.

6.8 E6F - Electro-optical Technology

6.8.1 Overview

The *Electro-optical Technology* question group in Subelement E6 tests you on the properties and operating characteristics of different optical devices such as photocells and display devices. The *Electro-optical Technology* group covers topics such as
- Photoconductivity
- Photovoltaic devices
- Optical sensors and encoders
- Optoisolation

The test producer will select one of the 11 questions in this group for your exam.

6.8.2   Questions

E6F01 What absorbs the energy from light falling on a photovoltaic cell?
A. Protons
B. Photons
C. Electrons
D. Holes

The electrons absorb the photons in the light, so **Answer C** is the correct choice.

E6F02 What happens to the conductivity of a photoconductive material when light shines on it?
A. It increases
B. It decreases
C. It stays the same
D. It becomes unstable

The conductivity increase mentioned in **Answer A** allows the photoconductive material to operate, making this the correct choice. Answer B is the opposite, and Answers C and D are to distract you.

E6F03 What is the most common configuration of an optoisolator or opto-coupler?
A. A lens and a photomultiplier
B. A frequency modulated helium-neon laser
C. An amplitude modulated helium-neon laser
D. An LED and a phototransistor

Answer D suggests using a LED and phototransistor, as in Figure 6.3, for an optoisolator or optocoupler, so that is the right choice. Answers B and C are interesting devices that have nothing to do with optocouplers, so they are incorrect. The device needs a light source, so Answer A is incomplete.

E6F04 What is the photovoltaic effect?
A. The conversion of voltage to current when exposed to light
B. The conversion of light to electrical energy
C. The conversion of electrical energy to mechanical energy
D. The tendency of a battery to discharge when exposed to light

The photovoltaic effect is the conversion of light into electrical energy, which makes **Answer B** the correct choice.

E6F05 Which describes an optical shaft encoder?
- A. A device that detects rotation of a control by interrupting a light source with a patterned wheel
- B. A device that measures the strength of a beam of light using analog to digital conversion
- C. A digital encryption device often used to encrypt spacecraft control signals
- D. A device for generating RTTY signals by means of a rotating light source

The shaft encoder interrupts a light source with a patterned wheel, so **Answer A** is the right choice. Answers B and C are interesting but not relevant here. Answer D is a silly distraction.

E6F06 Which of these materials is most commonly used to create photo-conductive devices?
- A. A crystalline semiconductor
- B. An ordinary metal
- C. A heavy metal
- D. A liquid semiconductor

Metals can demonstrate photoconductivity, but not as much as semiconductors, so Answers B and C are incorrect. Photoconductivity affects liquid semiconductors less than crystalline semiconductors, so **Answer A** is the best choice among those given.

E6F07 What is a solid-state relay?
- A. A relay using transistors to drive the relay coil
- B. A device that uses semiconductors to implement the functions of an electromechanical relay
- C. A mechanical relay that latches in the on or off state each time it is pulsed
- D. A passive delay line

A solid-state relay uses semiconductors to realize an electromechanical relay function, so **Answer B** is the right choice.

E6F08 Why are optoisolators often used in conjunction with solid-state circuits when switching 120 VAC?
- A. Optoisolators provide a low impedance link between a control circuit and a power circuit
- B. Optoisolators provide impedance matching between the control circuit and power circuit
- C. Optoisolators provide a very high degree of electrical isolation between a control circuit and the circuit being switched
- D. Optoisolators eliminate the effects of reflected light in the control circuit

When switching high voltages, one often needs circuit protection. Designers can use the optoisolator's electrical isolation, as described in **Answer C**, for this function, making it the right choice. The other statements are untrue.

E6F09 What is the efficiency of a photovoltaic cell?
 A. The output RF power divided by the input DC power
 B. Cost per kilowatt-hour generated
 C. The open-circuit voltage divided by the short-circuit current under full illumination
 D. The relative fraction of light that is converted to current

In photovoltaic cells, the efficiency is the percentage of the light energy converted to electrical energy, as in **Answer D**. The other choices are to distract you.

E6F10 What is the most common type of photovoltaic cell used for electrical power generation?
 A. Selenium
 B. Silicon
 C. Cadmium Sulfide
 D. Copper oxide

The silicon photovoltaic cells have the best price point, so they are the most common, and **Answer B** is the correct choice. Designers use cadmium sulfide in photodetectors but not in power cells. The others are to distract you.

E6F11 What is the approximate open-circuit voltage produced by a fully illuminated silicon photovoltaic cell?
 A. 0.1 V
 B. 0.5 V
 C. 1.5 V
 D. 12 V

Think in terms of a diode junction voltage. The approximate photovoltaic cell voltage is 0.5 V, so **Answer B** is the correct choice.

Chapter 7

E7 — PRACTICAL CIRCUITS

7.1 Introduction

The *Practical Circuits* subelement brings more depth to the radio circuits and components we have seen in the Technician and General Class studies as well as earlier in this Extra Class study. This subelement has questions on analog and digital circuit elements. This area includes amplifiers, filters, impedance matching circuits, modulation and demodulation circuits, and signal processing techniques. The *Practical Circuits* subelement has the following question groups:
  A. Digital circuits
  B. Amplifiers
  C. Filters and matching networks
  D. Power sources
  E. Modulation and demodulation
  F. Digital Signal Processing
  G. Active Filters and op-amp circuits
  H. Oscillators and signal sources
Subelement 7 will generate eight questions on the Extra examination.

7.2 Radio Engineering Concepts

Logic Functions We saw the schematic symbols for logic gates in Chapter 6. While we saw basic logic functions in the General Class study, there are several additional logic functions that you need to be familiar with for the Extra Class exam. The functions describe the output states for combinations of input states. Each function uses two inputs, either a 0 or a 1. Table 7.1 lists the outputs for the logic functions with these inputs. The table is for two inputs, and you can also extend it for more inputs.

Table 7.1: Combinational logic functions.

Input 1	Input 2	AND	NAND	OR	NOR	XOR	XNOR
0	0	0	1	0	1	0	1
0	1	0	1	1	0	1	0
1	0	0	1	1	0	1	0
1	1	1	0	1	0	0	1

Table 7.2: Characteristics of RF amplifiers

Class	Definition	Efficiency	Characteristics
A	Operate over full 360° input angle	25 % to 30 %	linear operation
AB	Two amplifiers; each operates over 180° input angle	50 % to 60 %	mostly linear operation
B	Two amplifiers operate over 180° input angle	up to 65 %	acceptable linearity
C	Operate near 90° input angle	up to 80 %	poor linearity
D	Operates by rapidly switching to achieve high efficiency and lower weight		

Amplifier Classes   As we saw in the General Class study guide, many types of amplifiers are available. As a review, Table 7.2 lists the standard amplifier classes and their characteristics that you need to know. The Class A amplifier is the most linear but least power efficient. The Class AB and Class B amplifiers use two amplification segments to cover the entire 360° carrier cycle.

Filters and Matching Networks   Filters and matching networks have different functions, but designers use similar circuit techniques to form each. First, we will look at filters and then the matching networks.

Designers designate filter types by which frequency components they pass through the circuit and which frequency components the circuit attenuates. We work exclusively in the frequency domain and not the time domain for filtering. Designers characterize the filter by its magnitude response and its phase response using the phasor notation encountered in Section 5.2. For our purposes here, we will look at the magnitude characteristics. Designers express the magnitude in terms of the amount of signal attenuation in dB units as a function of frequency. For example, a 0-dB attenuation means no attenuation,

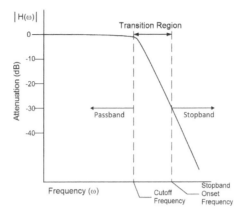

Figure 7.1: Key filter parameters in the frequency domain.

while a -20-dB attenuation means the signal is 100x smaller. Figure 7.1 illustrates several key filter characteristics that designers designate, including

Passband the frequency regions where the attenuation is less than 3 dB

Cutoff Frequency the attenuation is down 3 dB from the passband maximum (usually 0 dB)

Stopband the frequency region where the attenuation is at least a specified value, for example, -40 dB

Transition Region the frequency region between the passband and the stopband

Filter Order is related to the number of reactive elements in the circuit, and it determines the width of the transition region

Figure 7.2 illustrates the standard filter types:

Low Pass removes frequencies above the cutoff frequency

High Pass removes frequencies below the cutoff frequency

Band Pass passes frequencies in a defined region

Band Reject attenuates frequencies in a defined region; also called a "Notch Filter"

The actual circuits for the filter come from several design families. Figure 7.3 illustrates the amplitude response of several standard design families, including

Butterworth has the flattest magnitude response in the passband

Chebyshev has a rapid transition from the passband to the stopband but has a ripple in the passband response

Bessel has the best phase angle performance; good for digital signals

Elliptic has a very rapid transition from the passband to the stopband with ripple in the passband and the stopband; also known as a Cauer filter

Figure 7.3 normalizes the frequency axis to have the cutoff frequency at 1. The designer then scales the actual design to match the desired frequency region of interest. Figure 7.4 illustrates *LC* circuit designs for Low Pass Filter (LPF) and High Pass Filter (HPF) applications. The filter is the *LC* network between the

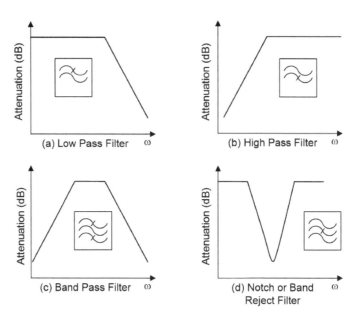

Figure 7.2: Frequency domain representation of standard filter types to modify signals.

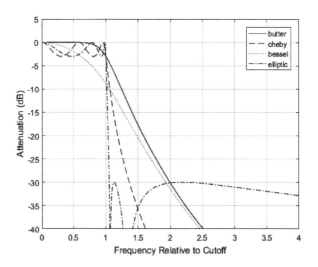

Figure 7.3: Examples of a Butterworth, Chebyshev, Bessel, and Elliptic filter magnitude response.

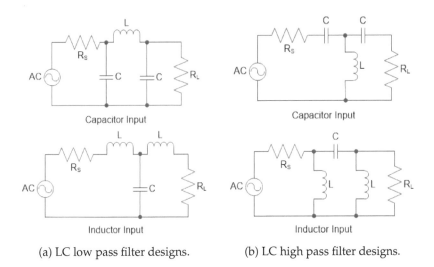

(a) LC low pass filter designs.   (b) LC high pass filter designs.

Figure 7.4: Low pass and high pass filter designs based on LC components.

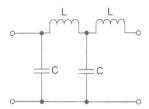

Figure 7.5: Pi-L matching network configuration.

source impedance, R_S, and the load impedance, R_L. The designer can realize each filter family by picking the appropriate inductor and capacitor values. The figure shows two approaches to each filter design: one with the capacitor as the first element and one with the inductor as the first element.

The designer can modify the LC networks for the LPF filters to match the impedance between the source and the load. The networks shown in Figure 7.4 are known as "Pi" networks when the capacitor is the first element because they look like the Greek letter π. Alternatively, if an inductor is the first element, designers call it a "T" network due to its shape. Figure 7.5 shows a slight modification to the Pi network as the "Pi-L" network with an inductor added to the output leg of the "Pi". Matching network design tools help designers select the LC components based on the matching requirements.

Oscillators Designers have several circuits to make an oscillator to produce sinusoidal waveforms at a specified frequency. The main requirement is to have multiple reactive elements to establish the variable signal. Figure 7.6 shows three common oscillator types: the Colpitts, the Hartley, and the Pierce. The

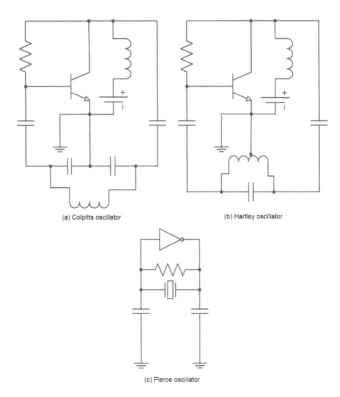

(a) Colpitts oscillator   (b) Hartley oscillator

(c) Pierce oscillator

Figure 7.6: Common oscillator configurations.

Colpitts and the Hartley oscillators have "tank circuits" composed of capacitors and inductors to set up the oscillation. The Colpitts design divides the single capacitor in the tank circuit into two capacitors with the variable signal exiting between the capacitors. The Hartley design replaces the split capacitors in the tank with a tapped inductor and a single capacitor where the inductor was. The transistors in the circuits are to provide signal gain. The Pierce oscillator uses a crystal in place of the inductor with the capacitors. The Colpitts and Hartley oscillators produce analog waveforms, while the Pierce oscillator has digital waves.

SSB Modulators   There are two methods for generating Single Sideband (SSB) modulation: a Hilbert Transform filter and a Band Pass Filter. Figure 7.7 shows block diagrams for these methods. The Hilbert Transform is a special filter with specific phase-shifting properties rather than amplitude attenuation properties. This method uses an oscillator's output and a 90° phase-shifted version going into two mixers. This process is known as *quadrature modulation*. The non-shifted branch is the in-phase (*I*) component, and the shifted branch is the quadrature-phase (*Q*) component. The Band Pass Filter (BPF) method specifically removes

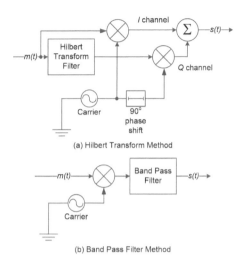

(a) Hilbert Transform Method

(b) Band Pass Filter Method

Figure 7.7: Methods for producing SSB transmissions.

the undesired sideband from a Dual Sideband - Suppressed Carrier (DSB-SC) Amplitude Modulation (AM) signal. The designer can realize these methods via hardware or software in a Software Defined Radio (SDR).

FM Systems   Frequency Modulation (FM) systems are usually more complicated than AM systems (the same is true for Phase Modulation (PM) systems too). There are multiple methods for generating FM. Two common approaches are the reactance modulator and the Phase Locked Loop (PLL) based modulator as in Figure 7.8. The figure shows a FM reactance modulator. Designers can use a similar circuit to generate PM. Notice that the reactance modulator uses a *LC* tank circuit configuration. The PLL circuit uses a Voltage Controlled Oscillator (VCO) to generate the carrier deviation in response to the message signal to make the transmitted FM signal.

The FM technique does not treat the message signal frequency contents equally. The higher-frequency components of the message signal will have a lower Signal-to-Noise Ratio (SNR) than the lower-frequency components with the FM system. To equalize this situation, designers use a "pre-emphasis" filter on the transmitter side to amplify the higher frequencies and a "de-emphasis" filter on the receiver side to bring them back in alignment. Figure 7.9 shows this placement.

PSK Modulators   Digital Phase Shift Keying (PSK) modulation is a limiting case of analog modulation systems, as Figure 7.10 illustrates. The digital input data may have pulse shaping filtering added for spectrum control (see more in Chapter 8). Designers use a DSB-SC modulator to make the transmitted signal. There are two quadrature modulation channels: the *I* and *Q* channels.

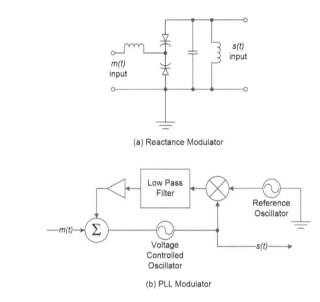

(a) Reactance Modulator

(b) PLL Modulator

Figure 7.8: Reactance and PLL FM modulators.

Figure 7.9: Pre-emphasis and de-emphasis filters in a FM system.

The circuit adds these signals together to make the output signal. Because the two channels are orthogonal, they do not interfere with each other. The demodulator uses a product detector with additional circuitry to recover the data in the receiver.

7.3 E7A - Digital Circuits

7.3.1 Overview

The *Digital Circuits* question group in Subelement E7 tests you on two-input logic functions and other digital devices like flip-flops. The *Digital Circuits* group covers topics such as

- Digital circuit principles and logic circuits
- Classes of logic elements
- Positive and negative logic
- Frequency dividers
- Truth tables

The test producer will select one of the 11 questions in this group for your exam.

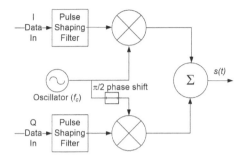

Figure 7.10: QPSK modulator formed from two AM modulators.

7.3.2 Questions

E7A01 Which circuit is bistable?
   A. An AND gate
   B. An OR gate
   C. A flip-flop
   D. A bipolar amplifier

The bistable circuit has two stable states, and it is just a fancy name for a flip-flop, so **Answer C** is the right choice. Answers A and B are individual logic gates. Answer D relates to an analog device and not digital logic.

E7A02 What is the function of a decade counter?
   A. It produces one output pulse for every 10 input pulses
   B. It decodes a decimal number for display on a seven-segment LED display
   C. It produces 10 output pulses for every input pulse
   D. It decodes a binary number for display on a seven-segment LED display

A decade counter produces one output pulse for every ten input pulses as indicated in **Answer A**, so this is the right choice. Designers call the devices listed in Answers B and D display drivers and not decade counters, so these are incorrect. Answer C is the opposite function.

E7A03 Which of the following can divide the frequency of a pulse train by 2?
   A. An XOR gate
   B. A flip-flop
   C. An OR gate
   D. A multiplexer

Designers use flip-flops as frequency dividers, so **Answer B** is correct. The other digital circuit options cannot divide the frequency, so they are incorrect.

E7A04 How many flip-flops are required to divide a signal frequency by 4?
   A. 1
   B. 2
   C. 4
   D. 8

Each flip-flop can divide a signal's frequency by 2, so we compute $4 \div 2 = 2$ as in **Answer B**. The other choices are to distract you.

E7A05 Which of the following is a circuit that continuously alternates between two states without an external clock?
   A. Monostable multivibrator
   B. J-K flip-flop
   C. T flip-flop
   D. Astable multivibrator

If you know Greek, you will recognize "a" as meaning not and "stable" as being stable, so the astable multivibrator goes between two unstable states, and **Answer D** is the correct choice. The next question covers the monostable multivibrator. The flip-flops need a clock and other circuit elements.

E7A06 What is a characteristic of a monostable multivibrator?
   A. It switches momentarily to the opposite binary state and then returns to its original state after a set time
   B. It produces a continuous square wave oscillating between 1 and 0
   C. It stores one bit of data in either a 0 or 1 state
   D. It maintains a constant output voltage, regardless of variations in the input voltage

Answer A gives an accurate description, so this is the right choice. Electrically, Answers B, C, and D are not correct descriptions, so they are not good choices.

E7A07 What logical operation does a NAND gate perform?
   A. It produces logic 0 at its output only when all inputs are logic 0
   B. It produces logic 1 at its output only when all inputs are logic 1
   C. It produces logic 0 at its output if some but not all inputs are logic 1
   D. It produces logic 0 at its output only when all inputs are logic 1

As Table 7.1 shows, a NAND gate produces a logic 0 only when all inputs are a logic 1 as in **Answer D**. Answer A is for an OR gate, and Answer B is for an AND gate. Answer C is a close, but not exactly, description of an XOR logic gate.

E7A08 What logical operation does an OR gate perform?
   A. It produces logic 1 at its output if any or all inputs are logic 1
   B. It produces logic 0 at its output if all inputs are logic 1
   C. It only produces logic 0 at its output when all inputs are logic 1
   D. It produces logic 1 at its output if all inputs are logic 0

As we see in Table 7.1, an OR gate produces an output of a logic 1 if any input is a logic 1, making **Answer A** the right choice.

E7A09 What logical operation is performed by an exclusive NOR gate?
   A. It produces logic 0 at its output only if all inputs are logic 0
   B. It produces logic 1 at its output only if all inputs are logic 1
   C. It produces logic 0 at its output if any single input is logic 1
   D. It produces logic 1 at its output if any single input is logic 1

From Table 7.1, a two-input exclusive NOR gate, the output will be a logic 0 if either, but not both, inputs are a logic 1 as in **Answer C**.

E7A10 What is a truth table?
   A. A table of logic symbols that indicate the high logic states of an op-amp
   B. A diagram showing logic states when the digital device output is true
   C. A list of inputs and corresponding outputs for a digital device
   D. A table of logic symbols that indicate the logic states of an op-amp

The truth table, as shown in Table 7.1, matches the description **Answer C**. Answer A is incorrect because designers use truth tables with digital logic and not op-amps. Answer B is incorrect because it only lists one of the possible outputs, but not all of them. Answer D is incorrect because it is technobabble.

E7A11 What type of logic defines "1" as a high voltage?
   A. Reverse Logic
   B. Assertive Logic
   C. Negative Logic
   D. Positive Logic

In positive logic, the device represents logic 1 by a high voltage level, so **Answer D** is the right choice. Negative logic uses a low voltage for logic 1. Answers A and B do not describe logic levels properly.

7.4   E7B - Amplifiers

7.4.1   Overview

The *Amplifiers* question group in Subelement E7 builds on the amplifier classes from the General Class study as well as Radio Frequency (RF) amplifier circuits. The *Amplifiers* group covers topics such as
- Class of operation
- Vacuum tube and solid-state circuits
- Distortion and intermodulation
- Spurious and parasitic suppression
- Microwave amplifiers
- Switching-type amplifiers

The test producer will select one of the 18 questions in this group for your exam.

7.4.2   Questions

E7B01 For what portion of the signal cycle does each active element in a push-pull Class AB amplifier conduct?
   A. More than 180 degrees but less than 360 degrees
   B. Exactly 180 degrees
   C. The entire cycle
   D. Less than 180 degrees

The Class AB amplifier conducts for more than 180°, but less than 360°, so **Answer A** is correct. The Class A amplifier conducts for the full cycle, Class B conducts for 180°, and Class C conducts for less than 180°.

E7B02 What is a Class D amplifier?
   A. A type of amplifier that uses switching technology to achieve high efficiency
   B. A low power amplifier that uses a differential amplifier for improved linearity
   C. An amplifier that uses drift-mode FETs for high efficiency
   D. A frequency doubling amplifier

The Class D amplifier is a high-efficiency amplifier using switching technology, making **Answer A** the correct choice. The other options are to distract you.

E7B03 Which of the following components form the output of a class D amplifier circuit?
   A. A low-pass filter to remove switching signal components
   B. A high-pass filter to compensate for low gain at low frequencies
   C. A matched load resistor to prevent damage by switching transients
   D. A temperature compensating load resistor to improve linearity

A switch will produce high-frequency harmonics, so a low-pass filter is necessary to remove them. This need makes **Answer A** the right choice and Answer B the wrong choice. Answers C and D are to distract you.

E7B04 Where on the load line of a Class A common emitter amplifier would bias normally be set?
  A. Approximately halfway between saturation and cutoff
  B. Where the load line intersects the voltage axis
  C. At a point where the bias resistor equals the load resistor
  D. At a point where the load line intersects the zero bias current curve

Operating the amplifier mid-way between saturation and cutoff gives the amplifier the largest range, so it is the preferred place to operate the Class A amplifier. **Answer A** is the best choice among those given.

E7B05 What can be done to prevent unwanted oscillations in an RF power amplifier?
  A. Tune the stage for maximum SWR
  B. Tune both the input and output for maximum power
  C. Install parasitic suppressors and/or neutralize the stage
  D. Use a phase inverter in the output filter

These oscillations are within the amplifier, so Answers A, B, and D are not good choices since they deal with connections to external equipment. Neutralization, in **Answer C**, is the technique to remove oscillations within the circuitry, so this is the best choice among those given.

E7B06 Which of the following amplifier types reduces even-order harmonics?
  A. Push-push
  B. Push-pull
  C. Class C
  D. Class AB

This question asks about a general rule of thumb. In this case, the push-pull amplifier of **Answer B** is the right choice. Answer A is to distract you with a similar name. The choice of the Class C and Class AB amplifier, by itself, will not prevent oscillations, so Answers C and D are not good choices.

E7B07 Which of the following is a likely result when a Class C amplifier is used to amplify a single-sideband phone signal?
  A. Reduced intermodulation products
  B. Increased overall intelligibility
  C. Signal inversion
  D. Signal distortion and excessive bandwidth

SSB needs linear generation and recovery for proper operation. Using a non-linear Class C amplifier can cause distortions, which makes **Answer D** correct. The distortions will decrease intelligibility. If the amplifier inverts the sidebands, you have other problems as well. Distorting a signal with a Class C amplifier will not decrease intermodulation products.

E7B08 How can an RF power amplifier be neutralized?
  A. By increasing the driving power
  B. By reducing the driving power
  C. By feeding a 180-degree out-of-phase portion of the output back to the input
  D. By feeding an in-phase component of the output back to the input

A negative, or 180° out-of-phase feedback signal, from the output to the input usually performs the neutralization operation. This configuration makes **Answer C** the right choice. Answers A and B are to distract you. Answer D is incorrect because it is a positive feedback solution that makes things worse.

E7B09 Which of the following describes how the loading and tuning capacitors are to be adjusted when tuning a vacuum tube RF power amplifier that employs a Pi-network output circuit?
  A. The loading capacitor is set to maximum capacitance and the tuning capacitor is adjusted for minimum allowable plate current
  B. The tuning capacitor is set to maximum capacitance and the loading capacitor is adjusted for minimum plate permissible current
  C. The loading capacitor is adjusted to minimum plate current while alternately adjusting the tuning capacitor for maximum allowable plate current
  D. The tuning capacitor is adjusted for minimum plate current, and the loading capacitor is adjusted for maximum permissible plate current

Unless you regularly perform this adjustment, you will probably need to memorize the answer. **Answer D** gives the correct procedure. The other choices are just incorrect permutations on the relative roles of the components.

E7B10 In Figure E7-1, what is the purpose of R1 and R2?
  A. Load resistors
  B. Voltage divider bias
  C. Self bias
  D. Feedback

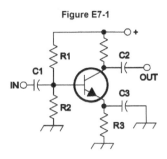

Figure E7-1

Figure 7.11: Figure E7-1.

Since the designer attached R1 and R2 to the transistor's input, not the output, they are not load resistors, so Answer A is incorrect. The two resistors form a voltage divider to set a fixed bias point for the transistor's base, so **Answer B** is the right choice here. They do not self-bias the transistor, so Answer C is incorrect. They do not link the output to the input, so they are not for feedback, and Answer D is also wrong.

E7B11 In Figure E7-1, what is the purpose of R3?
A. Fixed bias
B. Emitter bypass
C. Output load resistor
D. Self bias

As we saw above, R1 and R2 form the fixed bias for the transistor, so Answer A cannot be correct. The capacitor C3 is the emitter bypass capacitor, and this is not for emitter bypass (designers do not typically use resistors for bypass), so Answer B is incorrect. The resistor is not part of the output load, so Answer C is incorrect. It does perform self-bias functions by limiting the collector current, so **Answer D** is the correct choice.

E7B12 What type of amplifier circuit is shown in Figure E7-1?
A. Common base
B. Common collector
C. Common emitter
D. Emitter follower

Since the resistor grounds the emitter, it is a common-emitter configuration, and **Answer C** is the correct choice. The other options are to see if you know the amplifier configuration names.

E7B13 Which of the following describes an emitter follower (or common collector) amplifier?
A. A two-transistor amplifier with the emitters sharing a common bias resistor
B. A differential amplifier with both inputs fed to the emitter of the input transistor
C. An OR circuit with only one emitter used for output
D. An amplifier with a low impedance output that follows the base input voltage

An OR circuit is not an amplifier, so Answer C is a distraction. **Answer D** describes the correct configuration. Answers A and B are incorrect configurations.

E7B14 Why are switching amplifiers more efficient than linear amplifiers?
A. Switching amplifiers operate at higher voltages
B. The power transistor is at saturation or cut off most of the time
C. Linear amplifiers have high gain resulting in higher harmonic content
D. Switching amplifiers use push-pull circuits

The power transistor near saturation is the correct reason for the higher efficiency, which makes **Answer B** the right choice.

E7B15 What is one way to prevent thermal runaway in a bipolar transistor amplifier?
A. Neutralization
B. Select transistors with high beta
C. Use a resistor in series with the emitter
D. All of these choices are correct

Neutralization removes feedback, so Answer A is not a good choice for controlling thermal runaway. A high β will not prevent thermal problems, so Answer B is not correct either. Since A and B are incorrect, Answer D must also be wrong. An emitter resistor, as in **Answer C**, which controls current flow, and hence heating, is the right choice.

E7B16 What is the effect of intermodulation products in a linear power amplifier?
A. Transmission of spurious signals
B. Creation of parasitic oscillations
C. Low efficiency
D. All of these choices are correct

Because the linear amplifier does an excellent job of transmitting the components of the input signal, the intermodulation products appear as spurious signals, as suggested in **Answer A**. Answers B and C are not properties of linear amplification.

E7B17 Why are odd-order rather than even-order intermodulation distortion products of concern in linear power amplifiers?
A. Because they are relatively close in frequency to the desired signal
B. Because they are relatively far in frequency from the desired signal
C. Because they invert the sidebands causing distortion
D. Because they maintain the sidebands, thus causing multiple duplicate signals

The odd-order, especially the third-order, intermodulation distortion products appear close to the desired signal, which makes **Answer A** the correct choice for this question.

E7B18 What is a characteristic of a grounded-grid amplifier?
A. High power gain
B. High filament voltage
C. Low input impedance
D. Low bandwidth

The grounded-grid amplifier configuration is one found in many amateur rigs. One of its primary advantages, according to amplifier designers, is the low input impedance, so **Answer C** is the right choice among those given.

7.5   E7C - Filters and Matching Networks

7.5.1   Overview

The *Filters and Matching Networks* question group in Subelement E7 tests you on filter types and filter circuits. In addition to analog filters, we will also see digital filters. The *Filters and Matching Networks* group covers topics such as
- Types of networks
- Types of filters
- Filter applications
- Filter characteristics
- Impedance matching
- DSP filtering

The test producer will select one of the 12 questions in this group for your exam.

7.5.2   Questions

E7C01 How are the capacitors and inductors of a low-pass filter Pi-network arranged between the network's input and output?
A. Two inductors are in series between the input and output, and a capacitor is connected between the two inductors and ground
B. Two capacitors are in series between the input and output, and an inductor is connected between the two capacitors and ground
C. An inductor is connected between the input and ground, another inductor is connected between the output and ground, and a capacitor is connected between the input and output
D. A capacitor is connected between the input and ground, another capacitor is connected between the output and ground, and an inductor is connected between input and output

Figure 7.4a shows the arrangement for the low-pass Pi-network. This configuration makes **Answer D** the correct choice. Notice that the correct configuration looks like the Greek letter π. Remember: capacitors tend to pass high frequencies and block low frequencies, while inductors tend to pass low frequencies and block high frequencies. The Pi-network LPF shunts the high frequencies to ground while the low frequencies pass through to the filter's output.

E7C02 Which of the following is a property of a T-network with series capacitors and a parallel shunt inductor?
A. It is a low-pass filter
B. It is a band-pass filter
C. It is a high-pass filter
D. It is a notch filter

Figure 7.4b shows the T-network described in the question. This circuit is a form of a high-pass filter, in addition to transforming the impedance, so **Answer C** is the right choice.

E7C03 What advantage does a series-L Pi-L-network have over a series-L Pi-network for impedance matching between the final amplifier of a vacuum-tube transmitter and an antenna?
A. Greater harmonic suppression
B. Higher efficiency
C. Does not require a capacitor
D. Greater transformation range

Figure 7.5 shows a Pi-L network. The additional inductor on the output leg gives the matching network additional harmonic suppression. This result makes **Answer A** the correct choice.

E7C04 How does an impedance-matching circuit transform a complex impedance to a resistive impedance?
A. It introduces negative resistance to cancel the resistive part of impedance
B. It introduces transconductance to cancel the reactive part of impedance
C. It cancels the reactive part of the impedance and changes the resistive part to a desired value
D. Reactive currents are dissipated in matched resistances

If we go back to Figure 5.3, we can see that the pure resistances lie along the positive *x-axis* of the graph. A reactive component will displace the point above or below the *x-axis*. Therefore, you will need to introduce a reactive canceling element to make a resistive impedance from a reactive impedance. This process is the method of **Answer C**, so it is the correct choice for this question.

E7C05 Which filter type is described as having ripple in the passband and a sharp cutoff?
A. A Butterworth filter
B. An active LC filter
C. A passive op-amp filter
D. A Chebyshev filter

This question has the classic description of a Chebyshev filter, as shown in Figure 7.3, so **Answer D** is the right choice. The Butterworth filter is flat in the passband. Answers B and C do not describe any specific filter family, so they are incorrect.

E7C06 What are the distinguishing features of an elliptical filter?
A. Gradual passband rolloff with minimal stop band ripple
B. Extremely flat response over its pass band with gradually rounded stop band corners
C. Extremely sharp cutoff with one or more notches in the stop band
D. Gradual passband rolloff with extreme stop band ripple

You need to remember that an elliptical filter has an extremely sharp cutoff with deep notches in the stopband as described in **Answer C**. See Figure 7.3.

E7C07 Which describes a Pi-L-network used for matching a vacuum tube final amplifier to a 50-ohm unbalanced output?
A. A Phase Inverter Load network
B. A Pi-network with an additional series inductor on the output
C. A network with only three discrete parts
D. A matching network in which all components are isolated from ground

As we saw in Figure 7.5, a Pi-L network is a Pi-network with a series inductor on the output, so **Answer B** is the right choice. The others are distractions.

E7C08 Which of the following factors has the greatest effect on the bandwidth and response shape of a crystal ladder filter?
A. The relative frequencies of the individual crystals
B. The DC voltage applied to the quartz crystal
C. The gain of the RF stage preceding the filter
D. The amplitude of the signals passing through the filter

The crystal filter has the two crystals slightly offset in frequency, which sets the filter's overall bandwidth. This design makes **Answer A** the correct choice to answer this question. The Direct Current (DC) voltage will not be critical, so Answer B is incorrect. Answer C does not come into the design of the filter, so this is not a good choice. The amplitudes do not matter if the signal is above the noise floor, so Answer D is irrelevant.

E7C09 What is a crystal lattice filter?
- A. A power supply filter made with interlaced quartz crystals
- B. An audio filter made with four quartz crystals that resonate at 1 kHz intervals
- C. A filter using lattice-shaped quartz crystals for high-Q performance
- D. A filter with narrow bandwidth and steep skirts made using quartz crystals

Designers typically do not use a crystal lattice filter on power supplies. It does not have four crystals or resonate at 1 kHz intervals. The quartz crystals are not lattice-shaped. **Answer D** is the correct description.

E7C10 Which of the following filters would be the best choice for use in a 2 meter band repeater duplexer?
- A. A crystal filter
- B. A cavity filter
- C. A DSP filter
- D. An L-C filter

The key to this question is understanding that the question pool examiners are making a few assumptions about the duplexer configuration. The question assumes that the operator uses a single antenna for the repeater's input and output frequencies. The cavity filter gives the necessary isolation between the two frequencies for the duplexer, so **Answer B** is the correct answer. The other choices are not good filters to provide dual-frequency isolation.

E7C11 Which of the following describes a receiving filter's ability to reject signals occupying an adjacent channel?
- A. Passband ripple
- B. Phase response
- C. Shape factor
- D. Noise factor

The shape factor of **Answer C** controls how quickly the filter response drops off and how well it can reject other nearby signals. The other options do not control the filter's shape.

E7C12 What is one advantage of a Pi-matching network over an L-matching network consisting of a single inductor and a single capacitor?
- A. The Q of Pi-networks can be controlled
- B. L-networks cannot perform impedance transformation
- C. Pi-networks are more stable
- D. Pi-networks provide balanced input and output

Of the choices given, having a controllable Quality Factor (Q), as in **Answer A**, is the correct statement among the options given.

7.6 E7D - Power Sources

7.6.1 Overview

The *Power Sources* question group in Subelement E7 quizzes you on supplies and regulators for amateur operations. The *Power Sources* group covers topics such as

- Power supplies
- Voltage regulators
- Solar array charge controllers

The test producer will select one of the 15 questions in this group for your exam.

7.6.2 Questions

E7D01 How does a linear electronic voltage regulator work?
- A. It has a ramp voltage as its output
- B. It eliminates the need for a pass transistor
- C. The control element duty cycle is proportional to the line or load conditions
- D. The conduction of a control element is varied to maintain a constant output voltage

The constant output voltage in **Answer D** describes a linear voltage regulator, so that is the correct answer. Answer A is wrong because it would give a time-variable voltage output. Answers B and C are electrically incorrect statements.

E7D02 What is a characteristic of a switching electronic voltage regulator?
- A. The resistance of a control element is varied in direct proportion to the line voltage or load current
- B. It is generally less efficient than a linear regulator
- C. The controlled device's duty cycle is changed to produce a constant average output voltage
- D. It gives a ramp voltage at its output

A switching regulator has a switching circuit as described in **Answer C**, so this is the correct answer. Answer D would produce a time-variable output, so this is not a good choice. Answers A and B are incorrect statements.

E7D03 What device is typically used as a stable voltage reference in a linear voltage regulator?
- A. A Zener diode
- B. A tunnel diode
- C. An SCR
- D. A varactor diode

As we saw in Table 6.1, a Zener diode can provide a stable voltage reference, so **Answer A** is correct. Designers do not use the other diode types to provide stable reference choices, so they are incorrect choices.

E7D04 Which of the following types of linear voltage regulator usually make the most efficient use of the primary power source?
A. A series current source
B. A series regulator
C. A shunt regulator
D. A shunt current source

Answers A and D are not regulators, so they are not correct answers here. The correct answer is the series regulator in **Answer B**. The shunt regulator is for the next question.

E7D05 Which of the following types of linear voltage regulator places a constant load on the unregulated voltage source?
A. A constant current source
B. A series regulator
C. A shunt current source
D. A shunt regulator

Answers A and C do not correspond to voltage regulators, so they are incorrect. The correct choice is the shunt regulator in **Answer D**. The series regulator does not require a constant load.

Figure E7-2

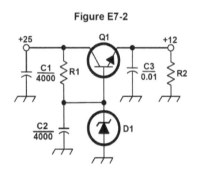

Figure 7.12: Figure E7-2.

E7D06 What is the purpose of Q1 in the circuit shown in Figure E7-2?
A. It provides negative feedback to improve regulation
B. It provides a constant load for the voltage source
C. It controls the current supplied to the load
D. It provides D1 with current

The transistor does not generate feedback, so Answer A is incorrect. The transistor does not connect to the output, so it cannot be a constant load (besides, it is the heart of the device), and Answer B is also incorrect. The transistor controls the current to flow, so **Answer C** is correct. It does not provide D1 with current, so Answer D is also a distraction for you.

E7D07 What is the purpose of C2 in the circuit shown in Figure E7-2?
A. It bypasses rectifier output ripple around D1
B. It is a brute force filter for the output
C. To self-resonate at the hum frequency
D. To provide fixed DC bias for Q1

The capacitor is to bypass any Alternating Current (AC) ripple, so **Answer A** is the correct choice. The capacitor does not connect to the output, so Answer B cannot be right. Answer C is not a good idea electrically, so this is not a good choice. Answer D is incorrect because the Zener diode sets the reference.

E7D08 What type of circuit is shown in Figure E7-2?
A. Switching voltage regulator
B. Grounded emitter amplifier
C. Linear voltage regulator
D. Monostable multivibrator

This example is not an amplifier circuit, so Answer B is incorrect. It is a linear voltage regulator making **Answer C** the right choice. Answer A is to see if you recognize the circuit. The multivibrator is from the digital logic family.

E7D09 What is the main reason to use a charge controller with a solar power system?
A. Prevention of battery undercharge
B. Control of electrolyte levels during battery discharge
C. Prevention of battery damage due to overcharge
D. Matching of day and night charge rates

The solar cells produce energy whenever the light hits them, which can damage the battery by overcharging. As **Answer C** says, the controller prevents this, so it is correct. Undercharging is not an issue. The controller will not help with electrolyte levels in a lead-acid battery. Answer D is a silly distraction because not much charging happens at night.

E7D10 What is the primary reason that a high-frequency switching type high-voltage power supply can be both less expensive and lighter in weight than a conventional power supply?
A. The inverter design does not require any output filtering
B. It uses a diode bridge rectifier for increased output
C. The high frequency inverter design uses much smaller transformers and filter components for an equivalent power output
D. It uses a large power factor compensation capacitor to create free power from the unused portion of the AC cycle

Smaller transformers and filter circuits will reduce weight and price, making

Answer C the correct choice. Answers A and B are not true electrically. Answer D is technobabble.

E7D11 What is the function of the pass transistor in a linear voltage regulator circuit?
  A. Permits a wide range of output voltage settings
  B. Provides a stable input impedance over a wide range of source voltage
  C. Maintains nearly constant output impedance over a wide range of load current
  D. Maintains nearly constant output voltage over a wide range of load current

The pass transistor is to provide the constant output voltage over varying load conditions as in **Answer D**. The others are distractions.

E7D12 What is the dropout voltage of an analog voltage regulator?
  A. Minimum input voltage for rated power dissipation
  B. Maximum amount that the output voltage drops when the input voltage is varied over its specified range
  C. Minimum input-to-output voltage required to maintain regulation
  D. Maximum that the output voltage may decrease at rated load

A linear voltage regulator needs a voltage difference across the device. Typically, there is a higher input voltage than the output voltage. The output disappears when the circuit does not provide the required voltage difference, so the regulator does not function. This result makes **Answer C** the correct choice.

E7D13 What is the equation for calculating power dissipation by a series linear voltage regulator?
  A. Input voltage multiplied by input current
  B. Input voltage divided by output current
  C. Voltage difference from input to output multiplied by output current
  D. Output voltage multiplied by output current

Here we go back to the power definition: $P = V \times I$. The I is the output current. Since we are looking for the dissipated power, we use the voltage difference between the input and the output as the V. **Answer C** uses this computation.

E7D14 What is the purpose of connecting equal-value resistors across power supply filter capacitors connected in series?
  A. Equalize the voltage across each capacitor
  B. Discharge the capacitors when voltage is removed
  C. Provide a minimum load on the supply
  D. All these choices are correct

Each purpose listed in Answers A, B, and C is correct, which makes **Answer D** the best choice.

E7D15 What is the purpose of a step-start circuit in a high voltage power supply?
  A. To provide a dual-voltage output for reduced power applications
  B. To compensate for variations of the incoming line voltage
  C. To allow for remote control of the power supply
  D. To allow the filter capacitors to charge gradually

A step-start circuit keeps capacitors from charging too quickly by limiting the inrush current at startup, so **Answer D** is the right choice.

7.7 E7E - Modulation and Demodulation

7.7.1 Overview

The *Modulation and Demodulation* question group in Subelement E7 tests you on analog modulation and demodulation circuits for radio communications. The *Modulation and Demodulation* group covers topics such as
  • Reactance, phase, and balanced modulators
  • Detectors
  • Mixer stages
The test producer will select one of the 11 questions in this group for your exam.

7.7.2 Questions

E7E01 Which of the following can be used to generate FM phone emissions?
  A. A balanced modulator on the audio amplifier
  B. A reactance modulator on the oscillator
  C. A reactance modulator on the final amplifier
  D. A balanced modulator on the oscillator

Designers use a balanced modulator with AM and not FM, so Answers A and D are incorrect. The reactance modulator of Figure 7.6 works with the oscillator, so **Answer B** is the right choice. Answer C is to distract you.

E7E02 What is the function of a reactance modulator?
  A. To produce PM signals by using an electrically variable resistance
  B. To produce AM signals by using an electrically variable inductance or capacitance
  C. To produce AM signals by using an electrically variable resistance
  D. To produce PM or FM signals by using an electrically variable inductance or capacitance

Remember: FM or PM and reactance modulators. This restriction means that we can eliminate Answers B and C since they deal with AM. Reactance deals with inductors and capacitors, not resistors. This relationship makes **Answer D** the right choice.

E7E03 What is a frequency discriminator stage in a FM receiver?
   A. An FM generator circuit
   B. A circuit for filtering two closely adjacent signals
   C. An automatic band-switching circuit
   D. A circuit for detecting FM signals

Designers use a discriminator in a FM receiver, so **Answer D** is the right choice here. It is not in the transmitter, so Answer A is incorrect. Electrically, it has nothing to do with Answers B and C, so these are not good choices.

E7E04 What is one way a single-sideband phone signal can be generated?
   A. By using a balanced modulator followed by a filter
   B. By using a reactance modulator followed by a mixer
   C. By using a loop modulator followed by a mixer
   D. By driving a product detector with a DSB signal

SSB is a form of AM, so using the balanced modulator and a filter like Figure 7.7(b) makes **Answer A** the right choice. Answer B is more for FM or PM generation, so it is incorrect here. Answer C is technobabble. Answer D is more for a receiver than a transmitter, so it is not a good choice.

E7E05 What circuit is added to an FM transmitter to boost the higher audio frequencies?
   A. A de-emphasis network
   B. A heterodyne suppressor
   C. A heterodyne enhancer
   D. A pre-emphasis network

This is the description of an FM pre-emphasis circuit used in Figure 7.9, so **Answer D** is the right choice. Answer A is the matching circuit in the receiver, so it is incorrect here. Answers B and C are technobabble.

E7E06 Why is de-emphasis commonly used in FM communications receivers?
   A. For compatibility with transmitters using phase modulation
   B. To reduce impulse noise reception
   C. For higher efficiency
   D. To remove third-order distortion products

Electrically, each statement in Answers B, C, and D is untrue. De-emphasis is to maintain compatibility with the transmitter, like that in Figure 7.9, which makes **Answer A** correct.

E7E07 What is meant by the term "baseband" in radio communications?
  A. The lowest frequency band that the transmitter or receiver covers
  B. The frequency range occupied by a message signal prior to modulation
  C. The unmodulated bandwidth of the transmitted signal
  D. The basic oscillator frequency in an FM transmitter that is multiplied to increase the deviation and carrier frequency

Baseband describes the unmodulated signal in its "natural form." These are the frequency components that go into the transmitter to modulate the carrier, making **Answer B** the correct choice. Each item in Answers A, C, and D is good to know, but they are not definitions of baseband.

E7E08 What are the principal frequencies that appear at the output of a mixer circuit?
  A. Two and four times the original frequency
  B. The square root of the product of input frequencies
  C. The two input frequencies along with their sum and difference frequencies
  D. 1.414 and 0.707 times the input frequency

A mixer's output is the two input frequencies and their sum and difference frequencies as in **Answer C**. The other options are not related to mixer operations.

E7E09 What occurs when an excessive amount of signal energy reaches a mixer circuit?
  A. Spurious mixer products are generated
  B. Mixer blanking occurs
  C. Automatic limiting occurs
  D. A beat frequency is generated

If the circuit provides a "hot" signal to the mixer input, then the mixer circuitry can generate spurious signals as in **Answer A**. The other options are not how mixers work.

E7E10 How does a diode envelope detector function?
  A. By rectification and filtering of RF signals
  B. By breakdown of the Zener voltage
  C. By mixing signals with noise in the transition region of the diode
  D. By sensing the change of reactance in the diode with respect to frequency

This detector type performs rectification and RF filtering, so **Answer A** is the right choice. Designers do not use Zener diodes in detection circuits, so Answer B is incorrect. Answer C is technobabble. Answer D is electrically incorrect.

E7E11 Which type of detector circuit is used for demodulating SSB signals?
A. Discriminator
B. Phase detector
C. Product detector
D. Phase comparator

Designers use a discriminator for FM, so Answer A is not a good choice. A product detector is a good method for recovering SSB, so **Answer C** is the best choice among those given.

7.8   E7F - Digital Signal Processing

7.8.1   Overview

The *Digital Signal Processing* question group in Subelement E7 tests you on the digital equivalents of the analog signal processing circuits for radio communications. The *Digital Signal Processing* group covers topics such as
• Filtering and other operations
• Software-defined radio fundamentals
• DSP modulation and demodulation
The test producer will select one of the 14 questions in this group for your exam.

7.8.2   Questions

E7F01 What is meant by direct digital conversion as applied to software defined radios?
A. Software is converted from source code to object code during operation of the receiver
B. Incoming RF is converted to a control voltage for a voltage controlled oscillator
C. Incoming RF is digitized by an analog-to-digital converter without being mixed with a local oscillator signal
D. A switching mixer is used to generate I and Q signals directly from the RF input

Answer A is a silly distraction to make you smile. The accurate description of direct conversion is to take the incoming RF, digitize it with an analog-to-digital converter, and then process the samples with a Digital Signal Processor (DSP) without using a mixer. **Answer C** is the correct description. Answers B and D are incorrect circuit operations.

E7F02 What kind of digital signal processing audio filter is used to remove unwanted noise from a received SSB signal?
  A. An adaptive filter
  B. A crystal-lattice filter
  C. A Hilbert-transform filter
  D. A phase-inverting filter

Designers can build an adaptive filter using DSP components, so **Answer A** is the best choice. A DSP cannot build a crystal filter. A Hilbert-transform filter is for processing SSB, not filtering noise. A phase-inversion filter will distort the signal and not remove the noise.

E7F03 What type of digital signal processing filter is used to generate an SSB signal?
  A. An adaptive filter
  B. A notch filter
  C. A Hilbert-transform filter
  D. An elliptical filter

If we wish to generate SSB, then the designer can realize the Hilbert-transform filter, like the one in Figure 7.7, using DSP components, and **Answer C** is the correct choice. Designers can use the adaptive, notch, and elliptical filters for noise filtering, but not modulation generation.

E7F04 What is a common method of generating an SSB signal using digital signal processing?
  A. Mixing products are converted to voltages and subtracted by adder circuits
  B. A frequency synthesizer removes the unwanted sidebands
  C. Varying quartz crystal characteristics emulated in digital form
  D. Signals are combined in quadrature phase relationship

As we saw in Figure 7.7, designers realize the Hilbert-transform method using a quadrature phase relationship, which makes **Answer D** the correct choice. The other options do not represent the actual circuit operations.

E7F05 How frequently must an analog signal be sampled by an analog-to-digital converter so that the signal can be accurately reproduced?
  A. At least half the rate of the highest frequency component of the signal
  B. At least twice the rate of the highest frequency component of the signal
  C. At the same rate as the highest frequency component of the signal
  D. At four times the rate of the highest frequency component of the signal

This question is asking about the Nyquist rate, which says the circuitry must sample the signal at least two times the rate of the highest frequency, as in **Answer B**. The other choices are to see if you know the Nyquist rate definition.

E7F06 What is the minimum number of bits required for an analog-to-digital converter to sample a signal with a range of 1 volt at a resolution of 1 millivolt?
- A. 4 bits
- B. 6 bits
- C. 8 bits
- D. 10 bits

A 1-V range sampled at a 1-mV resolution produces 1000 voltage levels. To represent this as a binary number requires 10 bits (1024 levels), so **Answer D** is the right choice. Answer A represents 16 levels, Answer B represents 64 levels, and Answer C represents 256 levels, so these are inadequate for the task.

E7F07 What function is performed by a Fast Fourier Transform?
- A. Converting analog signals to digital form
- B. Converting digital signals to analog form
- C. Converting digital signals from the time domain to the frequency domain
- D. Converting 8-bit data to 16-bit data

Fourier transforms are involved with going from the time domain to the frequency domain, so **Answer C** is the right choice. They are not involved with analog-to-digital conversion, so the other options are incorrect.

E7F08 What is the function of decimation?
- A. Converting data to binary code decimal form
- B. Reducing the effective sample rate by removing samples
- C. Attenuating the signal
- D. Removing unnecessary significant digits

In DSP, decimation refers to removing samples as in **Answer B**. The other options are to distract you.

E7F09 Why is an anti-aliasing digital filter required in a digital decimator?
- A. It removes high-frequency signal components that would otherwise be reproduced as lower frequency components
- B. It peaks the response of the decimator, improving bandwidth
- C. It removes low-frequency signal components to eliminate the need for DC restoration
- D. It notches out the sampling frequency to avoid sampling errors

Aliasing maps a higher frequency signal to a lower frequency. An anti-aliasing filter removes these components before sampling so that the Analog-to-Digital Converter (ADC) does not alias them. **Answer A** is the right choice. The other options will not prevent aliasing.

E7F10 What aspect of receiver analog-to-digital conversion determines the maximum receive bandwidth of a Direct Digital Conversion SDR?
A. Sample rate
B. Sample width in bits
C. Sample clock phase noise
D. Processor latency

The sample rate determines the bandwidth making **Answer A** the correct choice. Answers C and D are not relevant here. Answer B is how many bits represent each sample, which is not important here.

E7F11 What sets the minimum detectable signal level for a direct-sampling SDR receiver in the absence of atmospheric or thermal noise?
A. Sample clock phase noise
B. Reference voltage level and sample width in bits
C. Data storage transfer rate
D. Missing codes and jitter

The reference voltage and the voltage step size, which the sample width in bits determines, are essential for determining the minimum detectable signal level as in **Answer B**. The other options are not relevant for this determination.

E7F12 Which of the following is an advantage of a Finite Impulse Response (FIR) filter vs an Infinite Impulse Response (IIR) digital filter?
A. FIR filters can delay all frequency components of the signal by the same amount
B. FIR filters are easier to implement for a given set of passband rolloff requirements
C. FIR filters can respond faster to impulses
D. All these choices are correct

Delaying all frequency components by the same amount is needed for preserving the proper phase relationship, which makes **Answer A** the correct choice. The other options are not valid for Finite Impulse Response (FIR) filters.

E7F13 What is the function of taps in a digital signal processing filter?
A. To reduce excess signal pressure levels
B. Provide access for debugging software
C. Select the point at which baseband signals are generated
D. Provide incremental signal delays for filter algorithms

Taps provide incremental signal delays, as in **Answer D**. Answer A is a silly reason. The other options are distractions.

E7F14 Which of the following would allow a digital signal processing filter to create a sharper filter response?
- A. Higher data rate
- B. More taps
- C. Complex phasor representations
- D. Double-precision math routines

More taps permit better filter response, so **Answer B** is correct. A higher data rate will not give a sharper filter response, so Answer A is incorrect. Answers C and D are incorrect statements.

7.9   E7G - Active Filters and Op-amp Circuits

7.9.1   Overview

The *Active Filters and Op-amp Circuits* question group in Subelement E7 tests you on operational amplifiers used in signal processing circuits. The *Active Filters and Op-amp Circuits* group covers topics such as
- Active audio filters
- Device characteristics
- Basic circuit design
- Operational amplifiers

The test producer will select one of the 12 questions in this group for your exam.

7.9.2   Questions

E7G01 What is the typical output impedance of an op-amp?
- A. Very low
- B. Very high
- C. 100 ohms
- D. 1000 ohms

The op-amp has a "very low" output impedance, so **Answer A** is the correct choice. Be careful with "very high" since that is the right answer for the input, but not the output. The other answers are to distract you.

E7G02 What is ringing in a filter?
- A. An echo caused by a long time delay
- B. A reduction in high frequency response
- C. Partial cancellation of the signal over a range of frequencies
- D. Undesired oscillations added to the desired signal

Ringing in a circuit an unwanted oscillating signal in the filter as in **Answer D**. The other choices are inaccurate descriptions of filter ringing.

E7G03 What is the typical input impedance of an op-amp?
A. 100 ohms
B. 1000 ohms
C. Very low
D. Very high

The theoretical op-amp has an infinite input impedance. A good approximation to this is "very high," as in **Answer D**, so that is the correct choice here. As we saw, "very low" is the right answer for the output and not the input, so be careful here. The other answers are to try to distract you.

E7G04 What is meant by the "op-amp input offset voltage"?
A. The output voltage of the op-amp minus its input voltage
B. The difference between the output voltage of the op-amp and the input voltage required in the immediately following stage
C. The differential input voltage needed to bring the open loop output voltage to zero
D. The potential between the amplifier input terminals of the op-amp in an open loop condition

Since the op-amp is a differential amplifier based on the voltages at the input terminals, Answers B and C are the two choices to examine more carefully. The offset voltage is the input voltage to bring the output voltage to zero as in **Answer C**. The other options are not electrically correct.

E7G05 How can unwanted ringing and audio instability be prevented in an op-amp RC audio filter circuit?
A. Restrict both gain and Q
B. Restrict gain but increase Q
C. Restrict Q but increase gain
D. Increase both gain and Q

The correct solution is to restrict both the gain and the Q, so **Answer A** is the best choice to answer this question. Answers B, C, and D can each lead to ringing in the circuit by increasing either the Q or the gain, or both, so these are not good choices to answer the question.

E7G06 What is the gain-bandwidth of an operational amplifier?
A. The maximum frequency for a filter circuit using that type of amplifier
B. The frequency at which the open-loop gain of the amplifier equals one
C. The gain of the amplifier at a filter's cutoff frequency
D. The frequency at which the amplifier's offset voltage is zero

Designers define the op-amp gain-bandwidth metric as the frequency at which the open-loop gain equals 1 as in **Answer B**. The others are electronically incorrect.

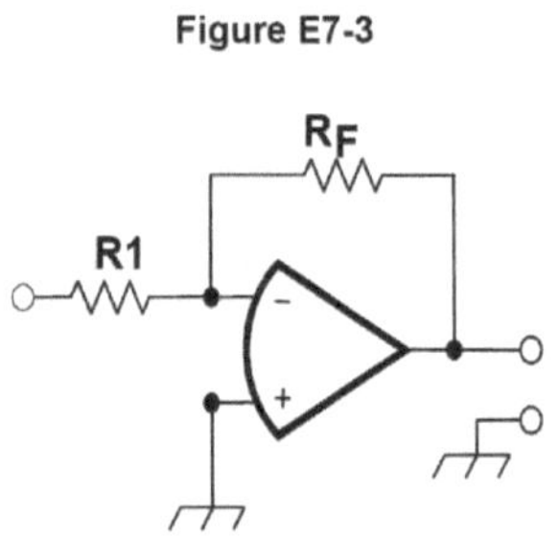

Figure E7-3

Figure 7.13: Figure E7-3.

E7G07 What magnitude of voltage gain can be expected from the circuit in Figure E7-3 when R1 is 10 ohms and RF is 470 ohms?
A. 0.21
B. 94
C. 47
D. 24

By performing circuit analysis on this circuit, and assuming a "perfect" amplifier, then the voltage gain is $V_{out}/V_{in} = -R_F/R1$. Substituting in the resistor values, we see that the magnitude of the voltage gain is $|G| = 470\,\Omega/10\,\Omega = 47$. The correct choice is **Answer C**. The other choices represent analysis mistakes.

E7G08 How does the gain of an ideal operational amplifier vary with frequency?
A. It increases linearly with increasing frequency
B. It decreases linearly with increasing frequency
C. It decreases logarithmically with increasing frequency
D. It does not vary with frequency

If the amplifier is "ideal", then there are no changes with frequency, so the correct choice is **Answer D**. Answers A, B, and C are to trick you.

E7G09 What will be the output voltage of the circuit shown in Figure E7-3 if R1 is 1000 ohms, RF is 10,000 ohms, and 0.23 volts DC is applied to the input?
A. 0.23 volts
B. 2.3 volts
C. -0.23 volts
D. -2.3 volts

Here we can use the voltage gain equation that we saw in the earlier question (without the magnitude sign). The output voltage is equal to $-R_F/R1$ times the input voltage. Using the numbers given here, $V_{out} = -(10\,000\,\Omega/1000\,\Omega)(0.23\,\text{V}) = -2.3\,\text{V}$. This computation makes **Answer D** the correct choice. Be careful here. Answer B has the right magnitude, but it is not the right polarity. Answers A and C have a factor of 10 missing.

E7G10 What absolute voltage gain can be expected from the circuit in Figure E7-3 when R1 is 1800 ohms and RF is 68 kilohms?
   A. 1
   B. 0.03
   C. 38
   D. 76

Here we use the gain magnitude analysis as before: $|G| = V_{out}/V_{in} = R_F/R1$. Using the resistor values given, we find $|G| = 68\,\text{k}\Omega/1800\,\Omega = 38$. We choose **Answer C** as the right choice for this question. The other choices represent analysis mistakes.

E7G11 What absolute voltage gain can be expected from the circuit in Figure E7-3 when R1 is 3300 ohms and RF is 47 kilohms?
   A. 28
   B. 14
   C. 7
   D. 0.07

Another question using the same analysis for the voltge gain magnitude: $|G| = R_F/R1$. Using the resistor values given, we find $|G| = 47\,\text{k}\Omega/3300\,\Omega = 14$. We choose **Answer B** as the right choice for this question. The other choices represent analysis mistakes.

E7G12 What is an operational amplifier?
   A. A high-gain, direct-coupled differential amplifier with very high input impedance and very low output impedance
   B. A digital audio amplifier whose characteristics are determined by components external to the amplifier
   C. An amplifier used to increase the average output of frequency modulated amateur signals to the legal limit
   D. A RF amplifier used in the UHF and microwave regions

The key item to spot here is "differential amplification" because that is the basis for the operational amplifier. This property makes **Answer A** the correct choice. Answer B is incorrect because the wording limits it to audio amplifiers. Answers C and D are not how the components work.

7.10   E7H - Oscillators and Signal Sources

7.10.1   Overview

The *Oscillators and Signal Sources* question group in Subelement E7 quizzes you on frequency generation circuits for radio communications. The *Oscillators and*

Signal Sources group covers topics such as
- Types of oscillators
- Synthesizers and phase-locked loops
- Direct digital synthesizers
- Stabilizing thermal drift
- Microphonics
- High-accuracy oscillators

The test producer will select one of the 15 questions in this group for your exam.

7.10.2   Questions

E7H01 What are three oscillator circuits used in amateur radio equipment?
- A. Taft, Pierce and negative feedback
- B. Pierce, Fenner and Beane
- C. Taft, Hartley and Pierce
- D. Colpitts, Hartley and Pierce

The Colpitts, Hartley, and Pierce oscillators, which Figure 7.6 shows, are the "big three" for amateur radio, so **Answer D** is the right choice to answer the question. Designers typically use the negative feedback mentioned in Answer A to damp oscillations, not to make an oscillator.

E7H02 What is a microphonic?
- A. An IC used for amplifying microphone signals
- B. Distortion caused by RF pickup on the microphone cable
- C. Changes in oscillator frequency due to mechanical vibration
- D. Excess loading of the microphone by an oscillator

Microphonics is the name for electronic devices transforming mechanical vibrations into electrical signals. **Answer C** is an application of this physical effect. Be careful because it is not part of a microphone.

E7H03 How is positive feedback supplied in a Hartley oscillator?
- A. Through a tapped coil
- B. Through a capacitive divider
- C. Through link coupling
- D. Through a neutralizing capacitor

Here we have three questions on the operation of the various oscillators. For the Hartley oscillator, the feedback is through a tapped coil, as in Figure 7.6, so **Answer A** is the right choice. The capacitive divider in Answer B is for the Colpitts oscillator, so it is incorrect here. The other two answers are to distract you.

E7H04 How is positive feedback supplied in a Colpitts oscillator?
  A. Through a tapped coil
  B. Through link coupling
  C. Through a capacitive divider
  D. Through a neutralizing capacitor

As we just saw, the tapped coil in Answer A is for the Hartley oscillator, so it is incorrect here. For a Colpitts oscillator, the feedback is through a capacitive divider, as in Figure 7.6, so **Answer C** is the right choice. The other two options are to distract you.

E7H05 How is positive feedback supplied in a Pierce oscillator?
  A. Through a tapped coil
  B. Through link coupling
  C. Through a neutralizing capacitor
  D. Through a quartz crystal

We know that the tapped coil is for the Hartley oscillator, so Answer A is incorrect. Answers B and C are still just distractions. From Figure 7.6, the correct choice is the quartz crystal of **Answer D**.

E7H06 Which of the following oscillator circuits are commonly used in VFOs?
  A. Pierce and Zener
  B. Colpitts and Hartley
  C. Armstrong and deForest
  D. Negative feedback and balanced feedback

There is a Zener diode, not a Zener oscillator, so Answer A is not a good choice. Armstrong and deForest are big names in FM, but not Variable Frequency Oscillators (VFOs), so Answer C is incorrect. Answer D is technobabble. The right choice is **Answer B**, but you probably guessed that.

E7H07 How can an oscillator's microphonic responses be reduced?
  A. Use of NP0 capacitors
  B. Reduce noise on the oscillator's power supply
  C. Increase the bias voltage
  D. Mechanically isolate the oscillator circuitry from its enclosure

Because microphonics are related to mechanical vibrations, mechanically isolating the oscillator, as in **Answer D**, is the right approach. The other options will not reduce mechanical influences.

E7H08 Which of the following components can be used to reduce thermal drift in crystal oscillators?
 A. NP0 capacitors
 B. Toroidal inductors
 C. Wirewound resistors
 D. Non-inductive resistors

Designers do not use inductors and resistors to reduce the thermal drift. The NP0 capacitors in **Answer A** are ceramic capacitors with high thermal stability, so they are the right components for this application.

E7H09 What type of frequency synthesizer circuit uses a phase accumulator, lookup table, digital to analog converter, and a low-pass anti-alias filter?
 A. A direct digital synthesizer
 B. A hybrid synthesizer
 C. A phase-locked loop synthesizer
 D. A diode-switching matrix synthesizer

A direct digital synthesizer uses these components, so **Answer A** is the right choice. Answer C uses VCO circuitry, so this is incorrect.

E7H10 What information is contained in the lookup table of a direct digital synthesizer (DDS)?
 A. The phase relationship between a reference oscillator and the output waveform
 B. Amplitude values that represent the desired waveform
 C. The phase relationship between a voltage-controlled oscillator and the output waveform
 D. Frequently used receiver and transmitter frequencies

A lookup table in memory tells the processor how to make the desired waveform, so **Answer B** is the right choice. Answer D might be helpful information, but it will not help to synthesize the waveform. The other answers are to distract you.

E7H11 What are the major spectral impurity components of direct digital synthesizers?
 A. Broadband noise
 B. Digital conversion noise
 C. Spurious signals at discrete frequencies
 D. Nyquist limit noise

Because of the approximations to the sine waves, harmonics at discrete frequencies arise, so **Answer C** is the right choice to answer this question. PLLs suffer from broadband noise, so Answer A is incorrect here. Nyquist limit noise is a distraction for you, so Answer D is incorrect.

E7H12 Which of the following must be done to insure that a crystal oscillator provides the frequency specified by the crystal manufacturer?
  A. Provide the crystal with a specified parallel inductance
  B. Provide the crystal with a specified parallel capacitance
  C. Bias the crystal at a specified voltage
  D. Bias the crystal at a specified current

The crystal needs the specified parallel capacitance as in **Answer B**. Be careful with Answer A because it has an inductor, which is incorrect. The others are not correct electrically.

E7H13 Which of the following is a technique for providing highly accurate and stable oscillators needed for microwave transmission and reception?
  A. Use a GPS signal reference
  B. Use a rubidium stabilized reference oscillator
  C. Use a temperature-controlled high Q dielectric resonator
  D. All of these choices are correct

Each approach in Answers A, B, and C can be part of the solution, so **Answer D** is the best choice for this question.

E7H14 What is a phase-locked loop circuit?
  A. An electronic servo loop consisting of a ratio detector, reactance modulator, and voltage-controlled oscillator
  B. An electronic circuit also known as a monostable multivibrator
  C. An electronic servo loop consisting of a phase detector, a low-pass filter, a voltage-controlled oscillator, and a stable reference oscillator
  D. An electronic circuit consisting of a precision push-pull amplifier with a differential input

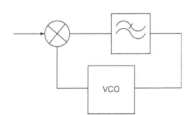

Figure 7.14 illustrates a PLL circuit. It has a phase detector, a low-pass filter, and a VCO. **Answer C** matches this configuration.

Figure 7.14: The block diagram for a PLL.

E7H15 Which of these functions can be performed by a phase-locked loop?
  A. Wide-band AF and RF power amplification
  B. Comparison of two digital input signals, digital pulse counter
  C. Photovoltaic conversion, optical coupling
  D. Frequency synthesis, FM demodulation

Because of the operational characteristics of a VCO, designers can use the PLL as both a frequency synthesizer and a FM demodulator. **Answer D** correctly describes the correct choices.

Chapter 8

E8 — SIGNALS AND EMISSIONS

8.1 Introduction

The *Signals and Emissions* subelement looks at signal representations in both the time domain and the frequency domain. We will examine signals as continuous and as sampled waveforms in the time domain. We will also further discuss radio circuits for modulation and demodulation. The *Signals and Emissions* subelement has the following question groups:

A. AC waveforms
B. Modulation and demodulation
C. Digital signals
D. Signal effects

Subelement 8 will generate four questions on the Extra examination.

8.2 Radio Engineering Concepts

Spectral Analysis   We can look at signals in two ways: in the time domain, like what we see on an oscilloscope, and in the frequency domain, like what we see on a spectrum analyzer. The *Fourier Transform* is the mathematical technique for going from one domain to the other. The Fourier Transform produces a complex function with real and imaginary components. We find working with this complex function convenient by expressing the result as a magnitude and phase. Engineers call the plot of the signal's magnitude versus frequency the *magnitude spectrum*, while they call a plot of the signal's phase versus frequency the *phase spectrum*. The magnitude spectrum tells us which frequency components have the strongest signal energy. Figure 8.1 shows the time domain and the magnitude spectrum for three periodic signals having a repetition

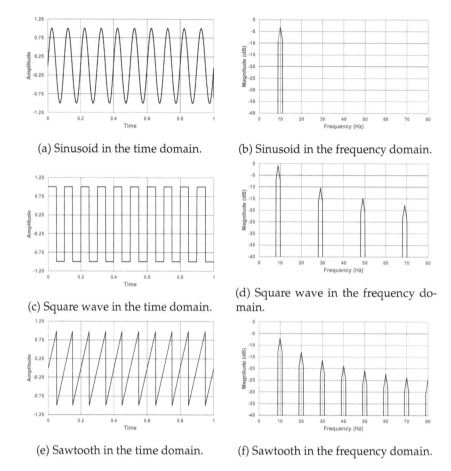

(a) Sinusoid in the time domain.

(b) Sinusoid in the frequency domain.

(c) Square wave in the time domain.

(d) Square wave in the frequency domain.

(e) Sawtooth in the time domain.

(f) Sawtooth in the frequency domain.

Figure 8.1: Periodic signals in the time and frequency domains. Each has a period of 10 Hz.

frequency of 10 Hz. These examples show that the magnitude spectrum has signal components only at multiples of the repetition frequency, also called the fundamental frequency. The sinusoid has a single frequency component. The square wave has the fundamental and odd-numbered multiples of it. The sawtooth has the fundamental and both even and odd multiples. A non-periodic signal such as a voice signal will have components at all frequencies.

Analog-to-Digital Conversion   Figure 4.2 showed the fundamental parts of an analog-to-digital conversion system. It consists of a Low Pass Filter (LPF) and an Analog-to-Digital Converter (ADC). The conversion process takes the signal input range and approximates it using fixed voltage levels as Figure 8.2 shows. The larger the number of levels, the finer the resolution of the input

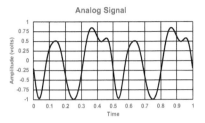
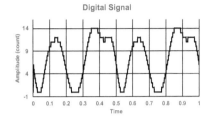

(a) Unsampled analog signal.   (b) Digital sampled signal.

Figure 8.2: Analog signal and the sampled digital representation.

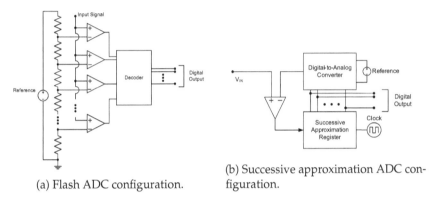

(a) Flash ADC configuration.

(b) Successive approximation ADC configuration.

Figure 8.3: Common analog-to-digital converters.

signal. If the ADC has N output bits, then the number of levels is $L = 2^N$.

There are many types of ADCs available on the market. For high-speed applications, such as digitizing signals in a Software Defined Radio (SDR), the designer needs a very-high speed ADC. The flash converter in Figure 8.3a is such a converter. It is based on a set of comparators that find the signal level in one clock cycle. For audio-type applications, a successive approximation ADC in Figure 8.3b has adequate speed and resolution. This ADC internally synthesizes the input signal over multiple clock cycles to generate the estimation.

The inverse of the analog-to-digital conversion process is the digital-to-analog conversion process. Figure 8.4 shows the components involved: a Digital-to-Analog Converter (DAC) to convert the digital code back to an analog voltage and a LPF to smooth the result and remove any higher-order harmonics.

FM Modulation Index and Deviation Ratio   As we saw in the General Class study guide, we can estimate the Frequency Modulation (FM) bandwidth by using Carson's rule. We estimate the transmission bandwidth, B, of a FM signal in terms of the carrier's maximum frequency deviation (the maximum

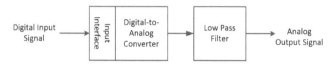

Figure 8.4: Digital-to-analog conversion elements.

difference the carrier moves from its unmodulated frequency), Δf, and the input message signal's bandwidth, M. Carson's rule estimates the transmission's bandwidth from $B = 2\,(\Delta f + M)$.

When the input signal is a pure sinusoidal tone, the signal bandwidth is the tone's frequency, f_m, and the transmission bandwidth is expressed using the transmitter's Modulation Index (β). We compute β from $\beta = \Delta f / f_m$. In this case, we estimate the transmission bandwidth from $B = 2\,(\beta + 1)\,f_m$.

When the input signal is a more complicated signal, like a voice or data signal, then the correct term is Deviation Ratio (D), which we compute from $D = \Delta f / M$. In practice, many call the deviation ratio the modulation index. With the Deviation Ratio, we estimate the transmission bandwidth from $B = 2\,(D + 1)\,M$.

Spectral Control   There are questions on the Extra Class examination concerning the necessary communications bandwidth. The *necessary bandwidth* is a regulatory term for how much frequency spectrum the transmitted signal should take with proper filtering and good design. In the previous paragraph, we saw the bandwidth, B, needed for FM. Regulators estimate the necessary bandwidth for other modulation formats as follows:

CW $B = KS$ where $3 < K < 5$ (4 is typical); S is the code sending rate

FM $B = 2\,(D + 1)\,M$; D is the deviation ratio and M is the message bandwidth

RTTY $B = S + 2K\Delta$; S is the sending character baud rate, Δ is the frequency shift, and $K = 1.2$ typically

PSK $B = 2R_b / \log_2 S$; R_b is the bit rate (bits/second) and S is the number of phase points (Binary Phase Shift Keying (BPSK) $S = 1$ and Quadrature Phase Shift Keying (QPSK) $S = 4$)

Notice: QPSK can send the same amount of data in half the bandwidth as BPSK. We can make similar computations for other modulation formats.

All the transmission modes require infinite transmission bandwidth in a mathematical sense. In addition to efficiently coding the transmitted symbols, we also need to control the transmitted bandwidth to permit channel sharing. One method digital transmission designers use achieves a narrow bandwidth by adding filters to shape the digital pulses before transmission. In Figure 7.10, we saw that we could add pulse shaping filters to the modulator. Figure 8.5 shows how filter choice affects the transmission bandwidth for a Phase Shift Keying (PSK) signal. The first panel of the figure shows the unfiltered PSK signal. The other panels show

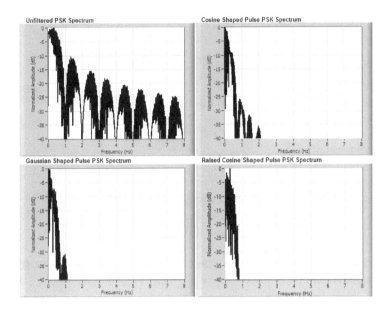

Figure 8.5: Effects of filter type on PSK transmission bandwidth.

Cosine Pulse Shape square waveform replaced with a cosine-shaped waveform

Gaussian Pulse Shape square waveform replaced with Gaussian function shape

Raised Cosine square waveform replaced with a special cosine shaping that minimizes interference between bits

In SDR-based systems, designers frequently execute these pulse shapes in software and not with *LC* filters. As the figure illustrates, square edges on the pulse shape require a large bandwidth for proper transmission. Using a rounded pulse shape, such as a cosine form, requires less bandwidth. This filtering means more users can share the channel.

8.3 E8A - AC Waveforms

8.3.1 Overview

The *AC Waveforms* question group in Subelement E8 looks at the analog and digital representations of signals. We will also examine the characteristics of those signals. The *AC Waveforms* group covers topics such as

- Fourier analysis; sine, square, sawtooth, and irregular waveforms
- AC measurements
- Average power and PEP of RF signals
- Analog-to-digital and digital-to-analog conversion
- Advantages of digital communications

The test producer will select one of the 11 questions in this group for your exam.

8.3.2 Questions

E8A01 What is the name of the process that shows that a square wave is made up of a sine wave plus all of its odd harmonics?
A. Fourier analysis
B. Vector analysis
C. Numerical analysis
D. Differential analysis

This question is asking about how we represent signals in the frequency domain. Fourier analysis, as in **Answer A**, is the means for this type of analysis. The other forms of analysis are helpful, but not for this application.

E8A02 Which of the following is a type of analog-to-digital conversion?
A. Successive approximation
B. Harmonic regeneration
C. Level shifting
D. Phase reversal

As we saw in Figure 8.3, there are two common types of ADCs: flash and successive approximation. **Answer A** is the correct choice, and the other options are not ADC types.

E8A03 What type of wave does a Fourier analysis show to be made up of sine waves of a given fundamental frequency plus all of its harmonics?
A. A sawtooth wave
B. A square wave
C. A sine wave
D. A cosine wave

Here we see how the waveforms will look on a spectrum analyzer that does the Fourier analysis. Applying a sawtooth wave to the input of a spectrum analyzer gives the graph in Figure 8.1f. Here, we see components at 10 Hz, 20 Hz, 30 Hz, 40 Hz, 50 Hz, 60 Hz, 70 Hz, and on out to infinity. Since these are both the even and the odd multiples of the fundamental frequency, engineers say all harmonics contribute to the sawtooth wave, so **Answer A** is correct. The square wave has just the odd harmonics, while the sine and cosine waves have just a single harmonic, so these choices are all incorrect.

E8A04 What is "dither" with respect to analog-to-digital converters?
A. An abnormal condition where the converter cannot settle on a value to represent the signal
B. A small amount of noise added to the input signal to allow more precise representation of a signal over time
C. An error caused by irregular quantization step size
D. A method of decimation by randomly skipping samples

The process of dithering adds a small amount of noise to aid in obtaining a more precise measurement, as in **Answer B**. The other options are effects in sampling signals, but engineers do not call them dithering.

E8A05 What of the following instruments would be the most accurate for measuring the RMS voltage of a complex waveform?
A. A grid dip meter
B. A D'Arsonval meter
C. An absorption wave meter
D. A true-RMS calculating meter

A test instrument specifically designed to measure the Root Mean Square (RMS) value is the best choice among those given, so we choose **Answer D**. A dip meter is better for measuring the resonant frequency, a D'Arsonval meter makes an instantaneous Direct Current (DC) current, voltage, or resistance measurement, and an absorption wavemeter is to distract you by trying to make you think of an absorption wattmeter.

E8A06 What is the approximate ratio of PEP-to-average power in a typical single-sideband phone signal?
A. 2.5 to 1
B. 25 to 1
C. 1 to 1
D. 100 to 1

This question addresses one of those "rules of thumb" that you may need to memorize. The Peak Envelope Power (PEP) will be larger than the average power for voice waveforms due to the shape of the voice waveform. This result means we can eliminate Answer C since that would make them equal. The correct choice is **Answer A** or 2.5:1. The other ratios are too large.

E8A07 What determines the PEP-to-average power ratio of a single-sideband phone signal?
  A. The frequency of the modulating signal
  B. Speech characteristics
  C. The degree of carrier suppression
  D. The amplifier gain

The characteristics of the voice waveform determine the PEP-to-average power ratio, so **Answer B** is the correct choice. The frequency of the modulating signal and the degree of carrier suppression are not used in the computation, while the amplifier gain comes after the transmitter modulates the carrier.

E8A08 Why would a direct or flash conversion analog-to-digital converter be useful for a software defined radio?
  A. Very low power consumption decreases frequency drift
  B. Immunity to out-of-sequence coding reduces spurious responses
  C. Very high speed allows digitizing high frequencies
  D. All of these choices are correct

A flash converter captures the analog-to-digital conversion quickly, so it is a good choice in SDR design, making **Answer C** correct. Answers A and B are false statements, which makes them, along with Answer D, incorrect.

E8A09 How many different input levels can be encoded by an analog-to-digital converter with 8-bit resolution?
  A. 8
  B. 8 multiplied by the gain of the input amplifier
  C. 256 divided by the gain of the input amplifier
  D. 256

An 8-bit converter can convert $2^8 = 256$ levels, which makes **Answer D** the correct choice. Answers B and C are technobabble. Answer A is a 3-bit resolution.

E8A10 What is the purpose of a low pass-filter used in conjunction with a digital-to-analog converter?
  A. Lower the input bandwidth to increase the effective resolution
  B. Improve accuracy by removing out-of-sequence codes from the input
  C. Remove harmonics from the output caused by the discrete analog levels generated
  D. All of these choices are correct

The LPF on the output of a DAC, as in Figure 8.4, is to interpolate between the samples and remove any high-frequency harmonics, which makes **Answer C** the best choice for this question. Answers A and B are incorrect statements, which makes Answer D incorrect.

E8A11 Which of the following is a measure of the quality of an analog-to-digital converter?
   A. Total harmonic distortion
   B. Peak envelope power
   C. Reciprocal mixing
   D. Power factor

Answers B, C, and D are not related to ADC performance, so they are eliminated. The total harmonic distortion of **Answer A** tells us the quality of the converter.

8.4 E8B - Modulation and Demodulation

8.4.1 Overview

The *Modulation and Demodulation* question group in Subelement E8 tests you on analog and digital radio techniques and the characteristics of the signals in the time and frequency domains. The *Modulation and Demodulation* group covers topics such as
   • Modulation methods
   • Modulation index and deviation ratio
   • Frequency and time-division multiplexing
   • Orthogonal Frequency Division Multiplexing
The test producer will select one of the 11 questions in this group for your exam.

8.4.2 Questions

E8B01 What is the modulation index of an FM signal?
   A. The ratio of frequency deviation to modulating signal frequency
   B. The ratio of modulating signal amplitude to frequency deviation
   C. The type of modulation used by the transmitter
   D. The bandwidth of the transmitted signal divided by the modulating signal frequency

For FM signals, the modulation index, β, is the ratio of the frequency deviation to the modulating signal frequency or $\beta = \Delta f / f_m$. This definition makes **Answer A** the correct choice. Answers B, C, and D are incorrect measurements to distract you.

E8B02 How does the modulation index of a phase-modulated emission vary with RF carrier frequency?
   A. It increases as the RF carrier frequency increases
   B. It decreases as the RF carrier frequency increases
   C. It varies with the square root of the RF carrier frequency
   D. It does not depend on the RF carrier frequency

Be careful with this tricky question. The modulation index is a function of the modulating signal, not of the specific carrier frequency, so **Answer D** is the correct choice. The others are distraction answers.

E8B03 What is the modulation index of an FM-phone signal having a maximum frequency deviation of 3000 Hz either side of the carrier frequency when the modulating frequency is 1000 Hz?
  A. 3
  B. 0.3
  C. 3000
  D. 1000

Applying the definition given earlier, we compute the modulation index as $\beta = 3\,\text{kHz}/1\,\text{kHz} = 3$. This result makes **Answer A** the correct choice. Answers B and C are off by factors of 10, so be careful when reading them.

E8B04 What is the modulation index of an FM-phone signal having a maximum carrier deviation of plus or minus 6 kHz when modulated with a 2 kHz modulating frequency?
  A. 6000
  B. 3
  C. 2000
  D. 1/3

Again, we compute the modulation index as $\beta = 6\,\text{kHz}/2\,\text{kHz} = 3$. This result makes **Answer B** the correct choice.

E8B05 What is the deviation ratio of an FM-phone signal having a maximum frequency swing of plus-or-minus 5 kHz when the maximum modulation frequency is 3 kHz?
  A. 60
  B. 0.167
  C. 0.6
  D. 1.67

Similar to β, we compute the deviation ratio as $D = 5\,\text{kHz}/3\,\text{kHz} = 1.67$. This computation makes **Answer D** the correct choice.

E8B06 What is the deviation ratio of an FM-phone signal having a maximum frequency swing of plus or minus 7.5 kHz when the maximum modulation frequency is 3.5 kHz?
  A. 2.14
  B. 0.214
  C. 0.47
  D. 47

You should be getting good at this by now. We compute the deviation ratio as $D = 7.5\,\text{kHz}/3.5\,\text{kHz} = 2.14$. This computation makes **Answer A** the correct choice.

E8B07 Orthogonal Frequency Division Multiplexing is a technique used for which type of amateur communication?
  A. High-speed digital modes
  B. Extremely low-power contacts
  C. EME
  D. OFDM signals are not allowed on amateur bands

Protocols use Orthogonal Frequency Division Multiplexing (OFDM) for high-speed digital modes, so **Answer A** is the right choice. Answer D is an incorrect statement. The WSJT-X and similar modes better serve the applications in Answers B and C.

E8B08 What describes Orthogonal Frequency Division Multiplexing?
  A. A frequency modulation technique that uses non-harmonically related frequencies
  B. A bandwidth compression technique using Fourier transforms
  C. A digital mode for narrow-band, slow-speed transmissions
  D. A digital modulation technique using subcarriers at frequencies chosen to avoid intersymbol interference

OFDM uses multiple subcarriers chosen to avoid intersymbol interference in the data transmission, making **Answer D** the correct choice. The other options are technically incorrect statements.

E8B09 What is deviation ratio?
  A. The ratio of the audio modulating frequency to the center carrier frequency
  B. The ratio of the maximum carrier frequency deviation to the highest audio modulating frequency
  C. The ratio of the carrier center frequency to the audio modulating frequency
  D. The ratio of the highest audio modulating frequency to the average audio modulating frequency

You should be able to spot **Answer B** as the correct one because it deals with carrier deviation and the input audio signal. Notice that none of the other choices deal with carrier deviation from its unmodulated frequency.

E8B10 What is frequency division multiplexing?
 A. The transmitted signal jumps from band to band at a predetermined rate
 B. Two or more information streams are merged into a baseband, which then modulates the transmitter
 C. The transmitted signal is divided into packets of information
 D. Two or more information streams are merged into a digital combiner, which then pulse position modulates the transmitter

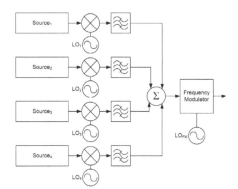

The graphic shows the Frequency Division Multiplexing (FDM) process: the designer merges the source signals to form a composite baseband signal for the input to the Radio Frequency (RF) transmitter. **Answer B** is the one with this description. Answer A may be a form of spread spectrum, while Answer C is packet communications.

Figure 8.6: Frequency division multiplexing channels for transmission.

E8B11 What is digital time division multiplexing?
 A. Two or more data streams are assigned to discrete sub-carriers on an FM transmitter
 B. Two or more signals are arranged to share discrete time slots of a data transmission
 C. Two or more data streams share the same channel by transmitting time of transmission as the sub-carrier
 D. Two or more signals are quadrature modulated to increase bandwidth efficiency

In Time Division Multiplexing (TDM), the transmitter gives each source a time slot for sending its data. **Answer B** is the correct choice to answer this question. Answer A is a form of FDM. Answer D is normal digital modulation like QPSK and Answer C is technobabble.

8.5 E8C - Digital Signals

8.5.1 Overview

The *Digital Signals* question group in Subelement E8 quizzes you on digital signal representations and methods for correcting transmission errors. The *Digital Signals* group covers topics such as
- Digital communication modes
- Information rate vs. bandwidth
- Error correction

The test producer will select one of the 12 questions in this group for your exam.

8.5.2 Questions

E8C01 How is Forward Error Correction implemented?
- A. By the receiving station repeating each block of three data characters
- B. By transmitting a special algorithm to the receiving station along with the data characters
- C. By transmitting extra data that may be used to detect and correct transmission errors
- D. By varying the frequency shift of the transmitted signal according to a predefined algorithm

Forward Error Correction (FEC) uses extra transmitted information to detect and correct transmission errors as in **Answer C**. Answer D is a form of FM, so it is incorrect here. Answers A and B are technobabble.

E8C02 What is the definition of symbol rate in a digital transmission?
- A. The number of control characters in a message packet
- B. The duration of each bit in a message sent over the air
- C. The rate at which the waveform changes to convey information
- D. The number of characters carried per second by the station-to-station link

The symbol rate is the rate at which the digital carrier waveform changes its state in response to the information as in **Answer C**. Be careful with Answer D because it can equal the symbol rate in some, but not all, circumstances. Answers A and B are important in digital communications, but not in this definition.

E8C03 Why should phase-shifting of a PSK signal be done at the zero crossing of the RF signal?
- A. To minimize bandwidth
- B. To simplify modulation
- C. It improves carrier suppression
- D. All of these choices are correct

Answers B and C are technically incorrect statements, making Answer D wrong. Maintaining a narrow bandwidth as is possible is good engineering practice, so **Answer A** is the right choice.

E8C04 What technique minimizes the bandwidth of a PSK31 signal?
A. Zero-sum character encoding
B. Reed-Solomon character encoding
C. Use of sinusoidal data pulses
D. Use of trapezoidal data pulses

As Figure 8.5 shows, the pulse shape will influence the transmitted bandwidth. Using a sinusoidal pulse shaping rather than sending square waves or trapezoidal waves will dramatically reduce transmitted bandwidth, and **Answer C** is the correct choice. Reed-Solomon is a form of FEC, which increases bandwidth. Answer A is a distraction.

E8C05 What is the approximate bandwidth of a 13-WPM International Morse Code transmission?
A. Approximately 13 Hz
B. Approximately 26 Hz
C. Approximately 52 Hz
D. Approximately 104 Hz

For Continuous Wave (CW), the channel bandwidth estimator computes the bandwidth, B, in terms of the sending speed, S, with $B = 4S = 4 \times 13 = 52$ Hz. **Answer C** is the correct choice.

E8C06 What is the bandwidth of a 170-hertz shift, 300-baud ASCII transmission?
A. 0.1 Hz
B. 0.3 kHz
C. 0.5 kHz
D. 1.0 kHz

In this case, the channel estimator computes the bandwidth, B, in terms of the character baud rate, S, and frequency shift, Δ, with $B = S + (1.2)(2\Delta) = 300 + (1.2)(170) = 504$ Hz. **Answer C** has the correct computation.

E8C07 What is the bandwidth of a 4800-Hz frequency shift, 9600-baud ASCII FM transmission?
A. 15.36 kHz
B. 9.6 kHz
C. 4.8 kHz
D. 5.76 kHz

In this case, the channel bandwidth estimator computes the bandwidth, B, in terms of the character baud, S, and frequency shift, Δ, with $B = S + (1.2)(2\Delta) = 9600 + (1.2)(4800) = 15.36\,\text{kHz}$. **Answer A** has the correct computation.

E8C08 How does ARQ accomplish error correction?
  A. Special binary codes provide automatic correction
  B. Special polynomial codes provide automatic correction
  C. If errors are detected, redundant data is substituted
  D. If errors are detected, a retransmission is requested

Automatic Repeat reQuest (ARQ) is an "Error control for data transmission in which the receiver detects transmission errors in a message and automatically requests a retransmission from the transmitter." (See http://www.atis.org/g lossary/). **Answer D** matches the standard definition.

E8C09 Which digital code allows only one bit to change between sequential code values?
  A. Binary Coded Decimal Code
  B. Extended Binary Coded Decimal Interchange Code
  C. Excess 3 code
  D. Gray code

The question is asking about Gray codes, so **Answer D** is the right choice. The other options are different types of coding.

E8C10 How may data rate be increased without increasing bandwidth?
  A. It is impossible
  B. Increasing analog-to-digital conversion resolution
  C. Using a more efficient digital code
  D. Using forward error correction

Generally, the bandwidth scales with the symbol rate for each digital modulation family, and the symbol rate is proportional to the data rate. A more efficient digital code family is required for a user to achieve an increased data rate without an associated bandwidth increase. For example, the user sends 2 bits per symbol rather than 1 bit per symbol. This technique makes **Answer C** the right choice. The other methods are not technically correct.

E8C11 What is the relationship between symbol rate and baud?
  A. They are the same
  B. Baud is twice the symbol rate
  C. Symbol rate is only used for packet-based modes
  D. Baud is only used for RTTY

The symbol rate and the baud are the same, so **Answer A** is the correct choice.

The other options are technically incorrect statements.

E8C12 What factors affect the bandwidth of a transmitted CW signal?
  A. IF bandwidth and Q
  B. Modulation index and output power
  C. Keying speed and shape factor (rise and fall time)
  D. All these choices are correct

Both the keying speed and shape factor affect the bandwidth for a CW signal making **Answer C** the correct choice. The other choices are not relevant in this determination.

8.6   E8D - Signal Effects

8.6.1   Overview

The *Signal Effects* question group in Subelement E8 examines the characteristics of transmitted analog and digital signals, including spread spectrum signals. The *Signal Effects* group covers topics such as
  • Keying defects and overmodulation of digital signals
  • Digital codes
  • Spread spectrum
The test producer will select one of the 11 questions in this group for your exam.

8.6.2   Questions

E8D01 Why are received spread spectrum signals resistant to interference?
  A. Signals not using the spread spectrum algorithm are suppressed in the receiver
  B. The high power used by a spread spectrum transmitter keeps its signal from being easily overpowered
  C. The receiver is always equipped with a digital blanker
  D. If interference is detected by the receiver it will signal the transmitter to change frequencies

Spread spectrum will tend to reject signals not aligned with its spreading code, so **Answer A** is the right choice. Answer B is incorrect because spread spectrum signals are power-limited on the amateur bands. Answer C does not describe what is happening technically. Answer D would make for a good receiver, but it does not explain how spread spectrum modulation works.

E8D02 What spread spectrum communications technique uses a high-speed binary bit stream to shift the phase of an RF carrier?
  A. Frequency hopping
  B. Direct sequence
  C. Binary phase-shift keying
  D. Phase compandored spread spectrum

This question is asking about the definition of Direct Sequence Spread Spectrum (DSSS), which makes **Answer B** the right choice. Frequency Hopping Spread Spectrum (FHSS) is the next question. Answer C is BPSK modulation with or without the spreading code. Answer D is a distraction.

E8D03 How does the spread spectrum technique of frequency hopping work?
  A. If interference is detected by the receiver it will signal the transmitter to change frequencies
  B. If interference is detected by the receiver it will signal the transmitter to wait until the frequency is clear
  C. A binary bit stream is used to shift the phase of an RF carrier very rapidly in a pseudorandom sequence
  D. The frequency of the transmitted signal is changed very rapidly according to a pseudorandom sequence also used by the receiving station

FHSS deals with changing the carrier frequency at the transmitter, and **Answer D** explains how it works. Answer A would be a nice feature if one could invent a popular-priced transceiver that works this way. The operators best execute Answer B, but it is not a transceiver feature. Answer C is DSSS, not FHSS.

E8D04 What is the primary effect of extremely short rise or fall time on a CW signal?
  A. More difficult to copy
  B. The generation of RF harmonics
  C. The generation of key clicks
  D. Limits data speed

Rapidly-changing on-off signals produce audio harmonics called key clicks, so **Answer C** is the right choice. The other options are to distract you.

E8D05 What is the most common method of reducing key clicks?
  A. Increase keying waveform rise and fall times
  B. Low-pass filters at the transmitter output
  C. Reduce keying waveform rise and fall times
  D. High-pass filters at the transmitter output

Increasing the rise and fall time will reduce the clicks, so **Answer A** is the right choice. Be careful with Answer C because it will make the problem worse. The filters in Answers B and D will not remove the clicks, so they are incorrect.

E8D06 What is the advantage of including parity bits in ASCII characters?
A. Faster transmission rate
B. The signal can overpower interfering signals
C. Foreign language characters can be sent
D. Some types of errors can be detected

Each ASCII character has a parity bit sent with the character. Receivers can use this parity bit to detect odd numbers (1 , 3, or 5) of errors, but not even numbers (2, 4, or 6). This property makes **Answer D** the correct choice. A single error is the most probable type of error in ASCII transmission, so this works well.

E8D07 What is a common cause of overmodulation of AFSK signals?
A. Excessive numbers of retries
B. Ground loops
C. Bit errors in the modem
D. Excessive transmit audio levels

The excessive audio levels of **Answer D** is the right choice. The other activities are potential problems, but not for this particular issue.

E8D08 What parameter evaluates distortion of an AFSK signal caused by excessive input audio levels?
A. Signal-to-noise ratio
B. Baud rate
C. Repeat Request Rate (RRR)
D. Intermodulation Distortion (IMD)

There will be Intermodulation Distortion (IMD) in the receiver if the transmitter distorts the signal by overmodulation. This definition makes **Answer D** the correct choice.

E8D09 What is considered an acceptable maximum IMD level for an idling PSK signal?
A. +10 dB
B. +15 dB
C. -20 dB
D. -30 dB

Having an IMD level as small as possible is desired. The -30 dB of **Answer D** is the best level of the options given since it is the smallest number.

E8D10 What are some of the differences between the Baudot digital code and ASCII?
  A. Baudot uses 4 data bits per character, ASCII uses 7 or 8; Baudot uses 1 character as a letters/figures shift code, ASCII has no letters/figures code
  B. Baudot uses 5 data bits per character, ASCII uses 7 or 8; Baudot uses 2 characters as letters/figures shift codes, ASCII has no letters/figures shift code
  C. Baudot uses 6 data bits per character, ASCII uses 7 or 8; Baudot has no letters/figures shift code, ASCII uses 2 letters/figures shift codes
  D. Baudot uses 7 data bits per character, ASCII uses 8; Baudot has no letters/figures shift code, ASCII uses 2 letters/figures shift codes

You need to remember Baudot 5 : ASCII 7 or 8. This property makes **Answer B** the only one that has the right number of bits. The others are incorrect variations on these code lengths.

E8D11 What is one advantage of using ASCII code for data communications?
  A. It includes built-in error correction features
  B. It contains fewer information bits per character than any other code
  C. It is possible to transmit both upper and lower case text
  D. It uses one character as a shift code to send numeric and special characters

Only **Answer C** is a correct statement among the choices given, so this is the right choice to answer this question. ASCII will detect errors but not correct them, so Answer A is incorrect. Answer B is wrong because ASCII has more bits than Baudot, not fewer. Answer D is not valid for ASCII.

Chapter 9

E9 — ANTENNAS AND TRANSMISSION LINES

9.1 Introduction

The *Antennas and Transmission Lines* subelement takes a "deep dive" into wire and directional antennas, matching, transmission lines, and their use in radio communications. We will also see more on antenna patterns, gain computations, radiated power, and the ever-present impedance matching. The *Antennas and Transmission Lines* subelement has the following question groups:

A. Basic antenna parameters
B. Antenna patterns and designs
C. Wire and phased array antennas
D. Directional antennas
E. Matching
F. Transmission lines
G. The Smith chart
H. Receiving antennas

Subelement 9 will generate eight questions on the Extra examination.

9.2 Radio Engineering Concepts

Antenna Patterns Each antenna radiates the Radio Frequency (RF) energy with different characteristics, depending upon the antenna type and manufacturing details. Figure 9.1 illustrates how designers describe the azimuthal energy distribution from the antenna. Think of this graphic as being a two-dimensional slice through a three-dimensional pattern. The third dimension occurs when you rotate the pattern around the y-axis. In this description, the relative gain appears as concentric rings, with the outermost ring being the

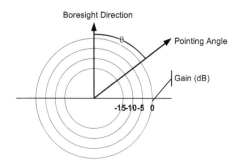

Figure 9.1: General formal for an azimuthal antenna pattern.

highest level. The plot specifies the gain level comparable to a reference antenna pattern. Engineers frequently determine the angular direction relative to the boresight, or direction the antenna is facing, along the y-axis.

Figure 9.2 illustrates several standard antenna patterns. These are

Isotropic — an ideal pattern where the antenna radiates uniformly in all directions; this is frequently the standard reference pattern

Aperture — a directional antenna such as a dish or patch antenna with a strongly preferred radiation direction

Dipole — a wire antenna usually mounted horizontally and above the ground so that the radiation has a wide distribution; this antenna is also frequently a standard reference pattern

Vertical — an antenna mounted perpendicular to the ground with most of the radiation off to the sides at a low angle relative to the horizon

Engineers assume the antenna radiates equally well in the transmitting as the receiving direction. These patterns are representative of what is known as the *far-field* pattern where the antenna works as designed. A good rule of thumb is that the far-field is the region beyond ten wavelengths from the antenna. Closer than that, the antenna is in the *near-field* area where the operation does not match the expected performance because interactions with the local environmental conditions strongly influence the electromagnetic fields.

When the reference antenna is an isotropic antenna, engineers usually write the gain as "dBi" for decibels - isotropic. Similarly, when the reference antenna, is a dipole, engineers usually write the gain as "dBd" for decibels - dipole. The gain of an isotropic antenna is 0 dB because the antenna emits all radiation uniformly over all directions. A dipole has a 2.15 dB gain relative to an isotropic antenna.

The Extra Class examination has several questions on how the antenna pattern of paired vertical antennas will look. The exact analysis is beyond the scope of this guide. However, Table 9.1 lists the information you need to know to answer the questions.

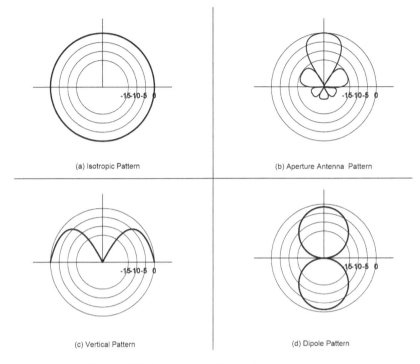

Figure 9.2: Representative antenna patterns.

Table 9.1: Vertical Antenna Pairs Radiation Patterns.

Spacing	Feed Phase	Pattern
½ wavelength	180°	Figure 8 along axis
½ wavelength	In phase	Figure 8 broadside
¼ wavelength	90°	Cardioid
¼ wavelength	180°	Figure 8 along axis
¼ wavelength	In phase	Elliptical
⅛ wavelength	180°	Figure 8 along axis

Antenna Parameters One of the key parameters for directional antennas is the Half Power Beam Width (HPBW) as illustrated in Figure 9.3. This measurement is the angular field where the gain is within 3 dB of the maximum level. Engineers consider this to be reasonably constant in practice. Figure 9.3 also shows the beamwidth between the first nulls or the place where that antenna pattern essential goes to zero output. It is like the HPBW and forms an alternative for describing how spatially confined the RF energy is. The smaller the antenna HPBW, the larger the gain. One cannot have an antenna with both a high gain

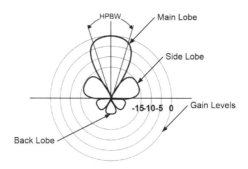

Figure 9.3: Antenna half-power beamwidth.

and a broad HPBW at the same time. To do so would imply that the antenna could radiate more RF energy than it receives at the input.

Engineers also characterize antennas by their *bandwidth*. The antenna's bandwidth rating is the frequency range where the Standing Wave Ratio (SWR) indicates a "good" match. A good rule of thumb is this is the region where the SWR is less than 2:1 to prevent damage to your transmitter. A related measurement is the Fractional Band Width (FBW) in terms of the upper, f_u, and lower, f_l, frequencies of the acceptable SWR range divided by the center frequency, f_c, and expressed as a percentage or

$$FBW = \frac{f_u - f_l}{f_c} \times 100\%$$

The antenna resistance is not just the simple ohmic resistance of the antenna. The total resistance of the antenna is the sum of the ohmic loss resistance and the *radiation resistance*. The radiation resistance is the feed point loss when radiating the RF energy. The *antenna efficiency* is the ratio of the radiation resistance to the total resistance.

Earlier, we saw the gain of dipole and isotropic antennas. Directional aperture antennas have gains that we can compute based on their clear aperture. For these antennas, we estimate the gain, G, in terms of the operating wavelength, λ, in meters or the operating frequency, f, in Hertz, and the speed of light, c, in meters per second by using the *universal gain equation*

$$G = \frac{4\pi A_e}{\lambda^2} = \frac{4\pi A_e f^2}{c^2}$$

A_e is the effective aperture of the antenna in square meters, which is related to the physical aperture (area), A_p, via the *aperture efficiency*, η_{ap}, by the equation

$$\eta_{ap} = \frac{A_e}{A_p}$$

Typically, aperture-type antennas have an aperture efficiency in the 45 % to 70 % range.

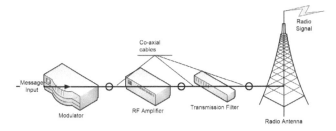

Figure 9.4: Components for computing the ERP. All element gains and losses need to be tracked.

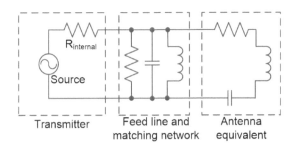

Figure 9.5: The equivalent circuit model for a communications system from the transmitter to the antenna.

Effective Radiated Power   The RF power generated at the transmitter is not necessarily the same as the power radiated by the antenna in a specific direction. Engineers use the Effective Radiated Power relative to a dipole (ERP) to compute the amount of RF power emitted by the system in a particular direction. This measurement is usually in the boresight direction for directional antennas unless specifically designated elsewhere. We use Figure 9.4 as a transmission system schematic. The total ERP is the sum of the transmitter output, amplifier gain, and antenna gain relative to a dipole minus any losses from cables, connectors, filters, etc., with all quantities and the result expressed in dB units. When engineers reference the ERP to an isotropic antenna, they frequently call it the Effective Isotropic Radiated Power (EIRP).

Antenna Matching   As we have progressed through the license examination process, we have seen many instances pointing out the need to have the antenna, the feed line, and the transmitter properly matched to enable the maximum RF power transmission through the system. Figure 9.5 shows a model of the process using Resistor-Inductor-Capacitor (RLC) equivalents for the components. The designer must match the RLC components for the specifics of the antenna's impedance. Figure 9.6 illustrates common matching approaches found in amateur systems.

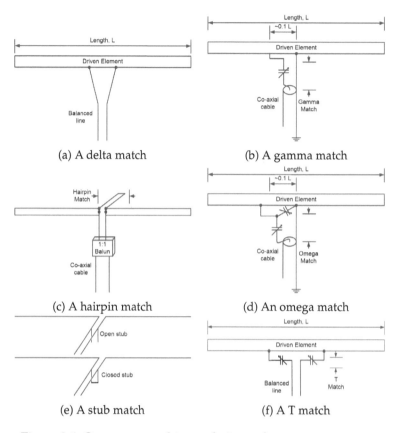

Figure 9.6: Common matching techniques for antenna systems.

Transmission Lines   Transmission lines are necessary for connecting transmitters and receivers to the antenna. In addition to the impedance, there are other important characteristics in practice. The first of these is the *velocity factor* for the transmission line. This measurement arises because the speed of the electromagnetic wave is slower in the transmission line than in free space. Physicists define the velocity factor as the ratio of the electromagnetic wave velocity to the free space velocity, and it is always less than one. Table 9.2 shows example velocity factors for various dielectric materials used in waveguides. The velocity factor implies that the electrical length of the transmission line is shorter than the physical length. This difference is essential when attempting to keep electromagnetic waves properly in phase. The electrical length equals the physical length multiplied by the velocity factor.

How the user terminates the transmission line will determine the inductive or capacitive nature of the line. The exact computation is beyond the scope of this study guide. The information in Table 9.3 is helpful in answering the Extra Class exam questions.

Table 9.2: Waveguide velocity factors.

Dielectric Material	Velocity Factor Range	Typical Velocity Factor
Air	0.95 – 0.99	0.99
Air Spaced Polyethylene	0.84	0.84
Foamed Polyethylene	0.73 – 0.88	0.80
Polyethylene	0.66 – 0.91	0.66
Teflon	0.695	0.695

Table 9.3: Waveguide Termination Characteristics.

Wavelength	Termination	Impedance
$\frac{1}{8}$ wavelength	shorted	inductive
$\frac{1}{8}$ wavelength	open	capacitive
$\frac{1}{4}$ wavelength	shorted	very high impedance
$\frac{1}{4}$ wavelength	open	very low impedance
$\frac{1}{2}$ wavelength	shorted	very low impedance
$\frac{1}{2}$ wavelength	open	very high impedance

9.3 E9A - Basic Antenna Parameters

9.3.1 Overview

The *Basic Antenna Parameters* question group in Subelement E9 tests you on antenna pattern shapes, antenna gain, transmitted power, and performance characteristics. The *Basic Antenna Parameters* group covers topics such as
- Radiation resistance
- Gain
- Beamwidth
- Efficiency
- Effective radiated power

The test producer will select one of the 13 questions in this group for your exam.

9.3.2 Questions

E9A01 What is an isotropic antenna?
A. A grounded antenna used to measure Earth conductivity
B. A horizontally polarized antenna used to compare Yagi antennas
C. A theoretical, omnidirectional antenna used as a reference for antenna gain
D. A spacecraft antenna used to direct signals toward the Earth

Here, the definition of "isotropic," uniformly radiating in all directions, does not help you select the right answer. The isotropic antenna does not exist in real life. It is more of a theoretical concept, and **Answer C** gives the correct usage. Answer A is technobabble. Designers compare the Yagi and other antennas with the isotropic at any polarization, so this is not as good a choice as Answer C. Engineers often call Answer D a "space-to-ground antenna" or an "Earth coverage antenna," which is not an isotropic antenna.

E9A02 What is the effective radiated power relative to a dipole of a repeater station with 150 watts transmitter power output, 2 dB feed line loss, 2.2 dB duplexer loss, and 7 dBd antenna gain?
   A. 1977 watts
   B. 78.7 watts
   C. 420 watts
   D. 286 watts

Working in dB units makes the computation easier. The net gain between the transmitter and the antenna output is $G = -2 - 2.2 + 7 = 2.80\,\text{dB}$. This represents a gain of $G = 10^{0.1*2.8} = 1.91$. The ERP is $P = 150\,\text{W} \times 1.91 = 286\,\text{W}$. This makes the correct choice **Answer D**.

E9A03 What is the radiation resistance of an antenna?
   A. The combined losses of the antenna elements and feed line
   B. The specific impedance of the antenna
   C. The value of a resistance that would dissipate the same amount of power as that radiated from an antenna
   D. The resistance in the atmosphere that an antenna must overcome to be able to radiate a signal

Only **Answer C** meets the technical definition, so it is correct. The combined losses are for computing the effective output power. Electrical resistance is important, but it is not radiation resistance. Answer D is a silly distraction.

E9A04 Which of the following factors affect the feed point impedance of an antenna?
   A. Transmission line length
   B. Antenna height
   C. The settings of an antenna tuner at the transmitter
   D. The input power level

The length of the antenna transmission line does not affect the intrinsic antenna feed point impedance, so Answer A is incorrect. Answers C and D are electrically incorrect, so they are not good choices. **Answer B** has the correct factor of antenna height, so this is the best choice among those given here.

E9A05 What is included in the total resistance of an antenna system?
A. Radiation resistance plus space impedance
B. Radiation resistance plus transmission resistance
C. Transmission-line resistance plus radiation resistance
D. Radiation resistance plus loss resistance

The radiation resistance and loss resistance, as listed in **Answer D**, make the total resistance. The impedance of free space is an actual physical concept, but it is not part of the radiation resistance. There is no transmission resistance, and the transmission line is not part of the antenna, so these are incorrect.

E9A06 What is the effective radiated power relative to a dipole of a repeater station with 200 watts transmitter power output, 4 dB feed line loss, 3.2 dB duplexer loss, 0.8 dB circulator loss, and 10 dBd antenna gain?
A. 317 watts
B. 2000 watts
C. 126 watts
D. 300 watts

The net gain between the transmitter and the antenna output is $G = -4 - 3.2 - 0.8 + 10 = 2$ dB. The gain is $G = 10^{0.1*2} = 1.58$. The ERP is $P = 200\,\text{W} \times 1.58 = 317\,\text{W}$. This makes the correct choice **Answer A** and the others distractions.

E9A07 What is the effective isotropic radiated power of a repeater station with 200 watts transmitter power output, 2 dB feed line loss, 2.8 dB duplexer loss, 1.2 dB circulator loss, and 7 dBi antenna gain?
A. 159 watts
B. 252 watts
C. 632 watts
D. 63.2 watts

The net gain between the transmitter and the antenna output is $G = -2 - 2.8 - 1.2 + 7 = 1$ dB. The gain is $G = 10^{0.1*1} = 1.26$. The EIRP is $P = 200\,\text{W} \times 1.26 = 252\,\text{W}$. This makes **Answer B** the correct choice and the others distractions.

E9A08 What is antenna bandwidth?
A. Antenna length divided by the number of elements
B. The frequency range over which an antenna satisfies a performance requirement
C. The angle between the half-power radiation points
D. The angle formed between two imaginary lines drawn through the element ends

Bandwidth is a frequency range measurement, so **Answer B** is the correct answer. Answers C and D cannot be correct because they deal with angles (beamwidth) and not frequencies (bandwidth). Answer A is technobabble.

E9A09 What is antenna efficiency?
A. Radiation resistance divided by transmission resistance
B. Radiation resistance divided by total resistance
C. Total resistance divided by radiation resistance
D. Effective radiated power divided by transmitter output

Engineers define the efficiency in terms of the radiation resistance and the total resistance, so **Answer B** is the correct formulation. There is no transmission resistance. Answer C has the order of the factors reversed, so be careful with this choice. Answer D is not how engineers compute antenna efficiency.

E9A10 Which of the following improves the efficiency of a ground-mounted quarter-wave vertical antenna?
A. Installing a radial system
B. Isolating the coax shield from ground
C. Shortening the radiating element
D. All these choices are correct

The best thing to improve a ground-mounted vertical antenna is to have a sound ground radial system, so **Answer A** is the best choice among those given. Answers C will change the tuning of the vertical antenna, so this is not a good choice. Answer B will make things worse by interrupting the system ground.

E9A11 Which of the following factors determines ground losses for a ground-mounted vertical antenna operating in the 3 MHz to 30 MHz range?
A. The standing wave ratio
B. Distance from the transmitter
C. Soil conductivity
D. Take-off angle

When you see ground losses, think "soil conductivity." The properties of the soil have a great deal to do with antenna operation, which makes **Answer C** the right choice for this question. The other factors are good to know but do not relate to this question.

E9A12 How much gain does an antenna have compared to a 1/2-wavelength dipole when it has 6 dB gain over an isotropic antenna?
A. 3.85 dB
B. 6.0 dB
C. 8.15 dB
D. 2.79 dB

A ½-wave dipole has a gain of 2.15 dB relative to an isotropic antenna, so the gain relative to the dipole is $6\,\text{dB} - 2.15\,\text{dB} = 3.85\,\text{dB}$. From this, **Answer A** is correct. Answer C adds the gain instead of subtracting it, while Answer B would be correct if the dipole had a gain of $0\,\text{dB}$ instead of $2.15\,\text{dB}$. Answer D is to distract you.

E9A13 What term describes station output, taking into account all gains and losses?
A. Power factor
B. Half-power bandwidth
C. Effective radiated power
D. Apparent power

This question is asking for the definition of Effective Radiated Power relative to a dipole (ERP), which makes **Answer C** the correct choice. The half-power bandwidth is the angle over which the antenna's loss is less than $3\,\text{dB}$. Earlier, we saw the definitions for the distraction answers power factor and apparent power.

9.4 E9B - Antenna Patterns and Designs

9.4.1 Overview

The *Antenna Patterns and Designs* question group in Subelement E9 tests you on antenna pattern measurements, modeling patterns with a computer, and the effects of the ground on the pattern. The *Antenna Patterns and Designs* group covers topics such as
- E and H plane patterns
- Gain as a function of pattern
- Antenna modeling

The test producer will select one of the 11 questions in this group for your exam.

9.4.2 Questions

E9B01 In the antenna radiation pattern shown in Figure E9-1, what is the beamwidth?
A. 75 degrees
B. 50 degrees
C. 25 degrees
D. 30 degrees

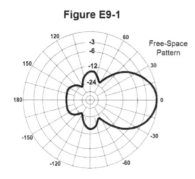

Figure E9-1

By looking at Figure 9.7, the pattern drops to 3 dB from maximum at about $\pm 25°$, so the beamwidth is about 50°. This result makes **Answer B** the correct choice. Answer C is one-half the beamwidth, so it is incorrect. Answers A and D are to distract you.

Figure 9.7: Figure E9-1.

E9B02 In the antenna radiation pattern shown in Figure E9-1, what is the front-to-back ratio?
A. 36 dB
B. 18 dB
C. 24 dB
D. 14 dB

The antenna gain at the front, 0°, is 0 dB while the gain at the back, 180°, is about half-way between -12 and -24 dB on this scale, so we will approximate it with -18 dB. The front-to-back ratio would be $G_{front} - G_{back} = 0\,dB - (-18\,dB) = 18\,dB$. This result makes **Answer B** the correct choice. Answer C and Answer D are for other angles.

E9B03 In the antenna radiation pattern shown in Figure E9-1, what is the front-to-side ratio?
A. 12 dB
B. 14 dB
C. 18 dB
D. 24 dB

The antenna gain at the front, 0°, is 0 dB while the gain at the side, 90° or 270°, is a bit less than -12 dB, so we will approximate it with -14 dB. The front-to-side ratio would be $G_{front} - G_{side} = 0$ decibel $- (-14\,dB) = 14\,dB$. This corresponds to the 14 dB in **Answer B**. Answer C and Answer D are for other angles.

E9B04 What is the front-to-back ratio of the radiation pattern shown in Figure E9-2?
A. 15 dB
B. 28 dB
C. 3 dB
D. 38 dB

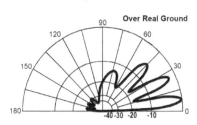

Figure E9-2

Figure 9.8: Figure E9-2.

The back lobes rise just above the -30-dB line, and the maximum is at the 0-dB line. These values result in a front-to-back ratio of 28 dB, making **Answer B** the right choice. The other answers are to distract you.

E9B05 What type of antenna pattern is shown in Figure E9-2?
A. Elevation
B. Azimuth
C. Radiation resistance
D. Polarization

This figure shows angles from $0°$ through $180°$, making it an elevation plot, and **Answer A** is the correct choice. An azimuthal pattern would go around a full circle. One cannot obtain radiation resistance or polarization from the radiation pattern.

E9B06 What is the elevation angle of peak response in the antenna radiation pattern shown in Figure E9-2?
A. 45 degrees
B. 75 degrees
C. 7.5 degrees
D. 25 degrees

By looking at the pattern plot, we can see the most prominent lobe occurs at $7.5°$, so that is the peak response. This identification makes **Answer C** the correct choice, and the others are to distract you.

E9B07 How does the total amount of radiation emitted by a directional gain antenna compare with the total amount of radiation emitted from a theoretical isotropic antenna, assuming each is driven by the same amount of power?
A. The total amount of radiation from the directional antenna is increased by the gain of the antenna
B. The total amount of radiation from the directional antenna is stronger by its front-to-back ratio
C. They are the same
D. The radiation from the isotropic antenna is 2.15 dB stronger than that from the directional antenna

If you think about it, the antenna cannot manufacture RF energy within it. All the antenna can do is either lose energy or direct it somewhere. The best an antenna can do is emit the same energy it receives regardless of antenna type. **Answer C** is correct because an ideal directional antenna radiates the same total energy as an isotropic antenna.

E9B08 What is the far field of an antenna?
- A. The region of the ionosphere where radiated power is not refracted
- B. The region where radiated power dissipates over a specified time period
- C. The region where radiated field strengths are constant
- D. The region where the shape of the antenna pattern is independent of distance

The far-field zone is where the antenna pattern's shape has stabilized and is now independent of distance, as listed in **Answer D**. This distance is about 10 wavelengths from the antenna. The other options are distractions.

E9B09 What type of computer program technique is commonly used for modeling antennas?
- A. Graphical analysis
- B. Method of Moments
- C. Mutual impedance analysis
- D. Calculus differentiation with respect to physical properties

Love your MoM! **Answer B**, the Method of Moments, is the usual method, so this is the right choice. Graphical analysis is too general, so it is not a good choice. Answer D is a form of analysis but not specific to antennas. Mutual impedance analysis is for transformers and not antennas.

E9B10 What is the principle of a Method of Moments analysis?
- A. A wire is modeled as a series of segments, each having a uniform value of current
- B. A wire is modeled as a single sine-wave current generator
- C. A wire is modeled as a single sine-wave voltage source
- D. A wire is modeled as a series of segments, each having a distinct value of voltage across it

The Method of Moments analysis technique uses wire segments and current values, so **Answer A** is the right choice. Answer B is technically incorrect. Answer C is meaningless without the current in the wire. Answer D is incorrect because it used voltage instead of current.

E9B11 What is a disadvantage of decreasing the number of wire segments in an antenna model below the guideline of 10 segments per half-wavelength?
A. Ground conductivity will not be accurately modeled
B. The resulting design will favor radiation of harmonic energy
C. The computed feed point impedance may be incorrect
D. The antenna will become mechanically unstable

One reason for decreasing the number of wire segments is that it improves the time required to run the analysis. The disadvantage is that the analysis is less precise. One way it may be less precise is in the computation of the feed point impedance, so **Answer C** is the best choice among those given.

9.5 E9C - Wire and Phased Array Antennas

9.5.1 Overview

The *Wire and Phased Array Antennas* question group in Subelement E9 quizzes you on various wire antennas and their performance. The *Wire and Phased Array Antennas* group covers topics such as
- Practical wire antennas
- Folded dipoles
- Phased arrays
- Ground effects near antennas

The test producer will select one of the 14 questions in this group for your exam.

9.5.2 Questions

E9C01 What is the radiation pattern of two 1/4-wavelength vertical antennas spaced 1/2-wavelength apart and fed 180 degrees out of phase?
A. Cardioid
B. Omni-directional
C. A figure-8 broadside to the axis of the array
D. A figure-8 oriented along the axis of the array

Making two antennas radiate in an omnidirectional pattern is very difficult, so Answer B is not a good choice. Unless you are an antenna guru, you will probably need to memorize the correct result for the next few questions. Table 9.1 may help. **Answer D** gives the correct choice for this configuration. Answers A and C correspond to other configurations.

E9C02 What is the radiation pattern of two 1/4-wavelength vertical antennas spaced 1/4-wavelength apart and fed 90 degrees out of phase?
- A. Cardioid
- B. A figure-8 end-fire along the axis of the array
- C. A figure-8 broadside to the axis of the array
- D. Omni-directional

This question is like the previous one, but now the phasing is 90° instead of 180°, and the spacing is ¼-wave instead of ½-wave. In this case, the correct choice is the cardioid pattern of **Answer A**. Answers B and C are for other configurations. The omnidirectional option is not a good one, as we indicated above.

E9C03 What is the radiation pattern of two 1/4-wavelength vertical antennas spaced a 1/2-wavelength apart and fed in phase?
- A. Omni-directional
- B. Cardioid
- C. A Figure-8 broadside to the axis of the array
- D. A Figure-8 end-fire along the axis of the array

We can use Table 9.1 again to find the correct answer. In this configuration, the right choice is the Figure-8 broadside pattern of **Answer C**. Answers B and D are for other arrangements. The omnidirectional option is still not a good one.

E9C04 What happens to the radiation pattern of an unterminated long wire antenna as the wire length is increased?
- A. The lobes become more perpendicular to the wire
- B. The lobes align more in the direction of the wire
- C. The vertical angle increases
- D. The front-to-back ratio decreases

In this case, the lobes align more in the direction of the wire, as in **Answer B**.

E9C05 Which of the following is a type of OCFD antenna?
- A. A dipole feed approximately 1/3 the way from one end with a 4:1 balun to provide multiband operation
- B. A remotely tunable dipole antenna using orthogonally controlled frequency diversity
- C. A folded dipole center-fed with 300-ohm transmission line
- D. A multiband dipole antenna using one-way circular polarization for frequency diversity

The user feeds the Off Center Fed Dipole (OCFD) antenna approximately 1/3 of the distance from one end and uses a balun to achieve multiband operation, as in **Answer A**. The other choices are all incorrect statements.

E9C06 What is the effect of a terminating resistor on a rhombic antenna?
  A. It reflects the standing waves on the antenna elements back to the transmitter
  B. It changes the radiation pattern from bidirectional to unidirectional
  C. It changes the radiation pattern from horizontal to vertical polarization
  D. It decreases the ground loss

If Answer A were true, it might damage the transmitter, so this is not a good choice. The resistor changes the antenna into a unidirectional pattern, as in **Answer B**. The resistor does not change the polarization state, so Answer C is not correct. Answer D is another distraction.

E9C07 What is the approximate feed point impedance at the center of a two-wire folded dipole antenna?
  A. 300 ohms
  B. 72 ohms
  C. 50 ohms
  D. 450 ohms

The correct choice in this instance is $300\,\Omega$, as in **Answer A**. Be careful with Answer C because many antennas are $50\,\Omega$, but not this one.

E9C08 What is a folded dipole antenna?
  A. A dipole one-quarter wavelength long
  B. A type of ground-plane antenna
  C. A half-wave dipole with an additional parallel wire connecting its two ends
  D. A dipole configured to provide forward gain

Answer A corresponds to a $\lambda/4$ dipole, which is incorrect. Answers B and D are untrue statements. The half-wave dipole with an additional parallel wire connecting its two ends in **Answer C** is the correct description.

E9C09 Which of the following describes a G5RV antenna?
  A. A multi-band dipole antenna fed with coax and a balun through a selected length of open wire transmission line
  B. A multi-band trap antenna
  C. A phased array antenna consisting of multiple loops
  D. A wide band dipole using shorted coaxial cable for the radiating elements and fed with a 4:1 balun

A G5RV antenna is another dipole-derivative antenna using an open-wire transmission line and a balun for multiband operation, as described in **Answer A**. The other options are incorrect choices.

E9C10 Which of the following describes a Zepp antenna?
A. A dipole constructed from zip cord
B. An end-fed dipole antenna
C. An omni-directional antenna commonly used for satellite communications
D. A vertical array capable of quickly changing the direction of maximum radiation by changing phasing lines

A Zepp antenna is an end-fed dipole, as in **Answer B**. Answer A should make you chuckle. The other descriptions are incorrect statements.

E9C11 How is the far-field elevation pattern of a vertically polarized antenna affected by being mounted over seawater versus soil?
A. The low-angle radiation decreases
B. Additional higher vertical angle lobes will appear
C. Fewer vertical angle lobes will be present
D. The low-angle radiation increases

The seawater will have the effect of increasing the low-angle radiation, so **Answer D** is the right choice. Answer A is the result for rocky soil, so this is incorrect. The other two options are to distract you.

E9C12 Which of the following describes an Extended Double Zepp antenna?
A. A wideband vertical antenna constructed from precisely tapered aluminum tubing
B. A portable antenna erected using two push support poles
C. A center-fed 1.25-wavelength antenna (two 5/8-wave elements in phase)
D. An end-fed folded dipole antenna

We saw the end-fed folded dipole as the answer to the normal Zepp above, so it is wrong here. The correct response is the 1.25-wavelength antenna of **Answer C**. The other choices are distractions.

E9C13 How does the radiation pattern of a horizontally polarized 3-element beam antenna vary with increasing height above ground?
A. The takeoff angle of the lowest elevation lobe increases
B. The takeoff angle of the lowest elevation lobe decreases
C. The horizontal beam width increases
D. The horizontal beam width decreases

As the height above the ground increases, the takeoff angle decreases. This property makes **Answer B** the correct choice.

E9C14 How does the performance of a horizontally polarized antenna mounted on the side of a hill compare with the same antenna mounted on flat ground?
  A. The main lobe takeoff angle increases in the downhill direction
  B. The main lobe takeoff angle decreases in the downhill direction
  C. The horizontal beamwidth decreases in the downhill direction
  D. The horizontal beamwidth increases in the uphill direction

Think of the downhill direction as providing extra height to the antenna. This effect results in the takeoff angle decreasing in that direction, making **Answer B** the right choice.

9.6 E9D - Directional Antennas

9.6.1 Overview

The *Directional Antennas* question group in Subelement E9 tests you on many aspects of directional antennas and their operational characteristics. The *Directional Antennas* group covers topics such as
  • Yagi antennas and parabolic reflectors
  • Circular polarization
  • Loading coils, top loading, and feed-point impedance of electrically short antennas
  • Antenna Q
  • RF grounding
The test producer will select one of the 12 questions in this group for your exam.

9.6.2 Questions

E9D01 How much does the gain of an ideal parabolic dish antenna change when the operating frequency is doubled?
  A. 2 dB
  B. 3 dB
  C. 4 dB
  D. 6 dB

As we saw in the Universal Gain Equation, the gain for a parabolic dish is proportional to the frequency squared, so if we double the frequency, we will raise the gain by four. In dB units, a factor of 4 is a gain of 6 dB. This result makes **Answer D** the correct choice for this question. The other options do not apply the dB conversion properly.

E9D02 How can linearly polarized Yagi antennas be used to produce circular polarization?
- A. Stack two Yagis fed 90 degrees out of phase to form an array with the respective elements in parallel planes
- B. Stack two Yagis fed in phase to form an array with the respective elements in parallel planes
- C. Arrange two Yagis perpendicular to each other with the driven elements at the same point on the boom fed 90 degrees out of phase
- D. Arrange two Yagis collinear to each other with the driven elements fed 180 degrees out of phase

Answer C has the correct configuration for making circular polarization with two Yagi antennas. Figure 9.9 shows what the layout looks like when used in a satellite tracking antenna for Very High Frequency (VHF) and Ultra High Frequency (UHF) bands. Be careful with **Answer D** because it looks similar to the right one, but it has the wrong phase for the feed. The other two choices are silly configurations to distract you.

Figure 9.9: Circularly polarized VHF and UHF crossed Yagi antennas.

E9D03 Where should a high Q loading coil be placed to minimize losses in a shortened vertical antenna?
- A. Near the center of the vertical radiator
- B. As low as possible on the vertical radiator
- C. As close to the transmitter as possible
- D. At a voltage node

In this case, the user should place the coil near the center, as in **Answer A**. The other answers will not work as intended.

E9D04 Why should an HF mobile antenna loading coil have a high ratio of reactance to resistance?
- A. To swamp out harmonics
- B. To lower the radiation angle
- C. To minimize losses
- D. To minimize the Q

Without knowing much about antennas, you might correctly pick minimizing the loss, as in **Answer C**, for your choice. The high reactance does not lower the radiation angle. Having a small Quality Factor (Q) is also not typically desired. Answer A is not necessarily true.

E9D05 What usually occurs if a Yagi antenna is designed solely for maximum forward gain?
  A. The front-to-back ratio increases
  B. The front-to-back ratio decreases
  C. The frequency response is widened over the whole frequency band
  D. The SWR is reduced

This question has another "rule-of-thumb". In this case, the front-to-back ratio decreases, as we would expect, if the designer optimizes the pattern for maximum forward gain. **Answer B** is the option with a correct statement. The other choices do not describe how the antenna behaves.

E9D06 What happens to the SWR bandwidth when one or more loading coils are used to resonate an electrically short antenna?
  A. It is increased
  B. It is decreased
  C. It is unchanged if the loading coil is located at the feed point
  D. It is unchanged if the loading coil is located at a voltage maximum point

The bandwidth will decrease if the user attaches loading coils, so **Answer B** is the correct choice. Answer A is the opposite of the right answer. The other options are to distract you.

E9D07 What is an advantage of using top loading in a shortened HF vertical antenna?
  A. Lower Q
  B. Greater structural strength
  C. Higher losses
  D. Improved radiation efficiency

By looking at the choices, one would hope that improving the radiation efficiency, as in **Answer D**, is the correct answer, and it is. Answer C would not be a good effect if it were true. Answer B is physically not true. Answer A is also not true.

E9D08 What happens as the Q of an antenna increases?
  A. SWR bandwidth increases
  B. SWR bandwidth decreases
  C. Gain is reduced
  D. More common-mode current is present on the feed line

If the designer increases the Q, the frequency region of low-SWR operation has a matching decrease, as in **Answer B**. Answer A is the exact opposite, so it is incorrect. Answers C and D are technically incorrect.

E9D09 What is the function of a loading coil used as part of an HF mobile antenna?
   A. To increase the SWR bandwidth
   B. To lower the losses
   C. To lower the Q
   D. To cancel capacitive reactance

A coil is inductive so that it can offset capacitive reactance, and **Answer D** is the right choice to answer this question. Answers A and B are consequences of balancing the capacitive reactance, so they are not as good a choice as Answer D. Answer C is to distract you.

E9D10 What happens to feed point impedance at the base of a fixed length HF mobile antenna when operated below its resonant frequency?
   A. The radiation resistance decreases and the capacitive reactance decreases
   B. The radiation resistance decreases and the capacitive reactance increases
   C. The radiation resistance increases and the capacitive reactance decreases
   D. The radiation resistance increases and the capacitive reactance increases

Answer B contains the description of what happens electrically, so this is the correct answer for this question. The other three answers are permutations to distract you.

E9D11 Which of the following conductors would be best for minimizing losses in a station's RF ground system?
   A. A resistive wire, such as spark plug wire
   B. A wide flat copper strap
   C. Stranded wire
   D. Solid wire

For a RF ground, you do not want a resistive wire, so Answer A is incorrect. A flat copper strap, as listed in **Answer B**, will have the necessary current carrying capacity and minimal loss, so this is the best choice among those given. The other wires are not as effective as the copper strap.

E9D12 Which of the following would provide the best RF ground for your station?
A. A 50-ohm resistor connected to ground
B. An electrically short connection to a metal water pipe
C. An electrically short connection to 3 or 4 interconnected ground rods driven into the Earth
D. An electrically short connection to 3 or 4 interconnected ground rods via a series RF choke

A sound ground system will have a firm, guaranteed, direct connection to Earth ground. The way to guarantee this is with ground rods directly driven into the Earth as given in **Answer C**. Resistors and chokes are not enhancers to this method, so Answers A and D are not good choices. Users frequently mention a metal water pipe for grounding, but, generally, it is not guaranteed to connect with the Earth ground. Therefore, we do not prefer it over a direct ground rod system.

9.7 E9E - Matching

9.7.1 Overview

The *Matching* question group in Subelement E9 quizzes you on matching network designs for antennas and feed lines. The *Matching* group covers topics such as
- Matching antennas to feed lines
- Phasing lines
- Power dividers

The test producer will select one of the 11 questions in this group for your exam.

9.7.2 Questions

E9E01 What system matches a higher-impedance transmission line to a lower-impedance antenna by connecting the line to the driven element in two places spaced a fraction of a wavelength each side of element center?
A. The gamma matching system
B. The delta matching system
C. The omega matching system
D. The stub matching system

This question has the description of the delta match network shown in Figure 9.6a, so **Answer B** is the correct choice. A gamma match, from Figure 9.6b, is for an unbalanced feed line. A stub, from Figure 9.6e, is a short section of a transmission line. The omega match, from Figure 9.6d, uses two capacitors.

E9E02 What is the name of an antenna matching system that matches an unbalanced feed line to an antenna by feeding the driven element both at the center of the element and at a fraction of a wavelength to one side of center?
   A. The gamma match
   B. The delta match
   C. The epsilon match
   D. The stub match

This question has the description of the gamma match from Figure 9.6b, so **Answer A** is the correct choice. The delta match is for high-impedance lines. A stub is a short section of a transmission line. The epsilon match is a distraction.

E9E03 What is the name of the matching system that uses a section of transmission line connected in parallel with the feed line at or near the feed point?
   A. The gamma match
   B. The delta match
   C. The omega match
   D. The stub match

This configuration describes the stub matching approach in Figure 9.6e, so **Answer D** is the correct choice. A gamma match is for an unbalanced feed line. The delta match is for high-impedance lines. The omega match uses capacitors.

E9E04 What is the purpose of the series capacitor in a gamma-type antenna matching network?
   A. To provide DC isolation between the feed line and the antenna
   B. To cancel the inductive reactance of the matching network
   C. To provide a rejection notch that prevents the radiation of harmonics
   D. To transform the antenna impedance to a higher value

Electrically, the designer uses a capacitor to cancel out the inductive reactance in a matching network, so **Answer B** is the proper choice to answer this question.

E9E05 How must an antenna's driven element be tuned to use a hairpin matching system?
   A. The driven element reactance must be capacitive
   B. The driven element reactance must be inductive
   C. The driven element resonance must be lower than the operating frequency
   D. The driven element radiation resistance must be higher than the characteristic impedance of the transmission line

For this type of match to work, the designer must tune the antenna with a net capacitive reactance to offset the hairpin's inductance. This property makes **Answer A** the right choice. Answer B is incorrect because the reactance is inductive rather than capacitive. Answer C and D are distractions.

E9E06 Which of these feed line impedances would be suitable for constructing a quarter-wave Q-section for matching a 100-ohm loop to 50-ohm feed line?
  A. 50 ohms
  B. 62 ohms
  C. 75 ohms
  D. 450 ohms

If uncorrected, we would have a 2:1 SWR problem in this system. To make the match, we need an impedance to split the difference or choose the 75 Ω impedance of **Answer C**. The other choices will not make the match as well.

E9E07 What parameter describes the interactions at the load end of a mismatched transmission line?
  A. Characteristic impedance
  B. Reflection coefficient
  C. Velocity factor
  D. Dielectric constant

The mismatched line will have reflections at the end, so the reflection coefficient makes the best description of what is happening electrically, and **Answer B** is the right choice. The others are variables to characterize the transmission line as a whole, but they will not directly tell you about electrical reflections.

E9E08 What is a use for a Wilkinson divider?
  A. It divides the operating frequency of a transmitter signal so it can be used on a lower frequency band
  B. It is used to feed high-impedance antennas from a low-impedance source
  C. It is used to divide power equally between two 50-ohm loads while maintaining 50-ohm input impedance
  D. It is used to feed low-impedance loads from a high-impedance source

The Wilkinson divider is a device that one can use as either a RF power splitter or a RF power combiner. As a power divider, the design isolates the output ports from each other, so the 50 Ω impedance is maintained, as indicated in **Answer C**.

E9E09 Which of the following is used to shunt-feed a grounded tower at its base?
  A. Double-bazooka match
  B. Hairpin match
  C. Gamma match
  D. All of these choices are correct

This question makes more sense with operational experience. For this case, you need to remember that the gamma match of **Answer C** is the right choice.

E9E10 Which of these choices is an effective way to match an antenna with a 100-ohm feed point impedance to a 50-ohm coaxial cable feed line?
- A. Connect a 1/4-wavelength open stub of 300-ohm twinlead in parallel with the coaxial feed line where it connects to the antenna
- B. Insert a 1/2-wavelength piece of 300-ohm twinlead in series between the antenna terminals and the 50-ohm feed cable
- C. Insert a 1/4-wavelength piece of 75-ohm coaxial cable transmission line in series between the antenna terminals and the 50-ohm feed cable
- D. Connect 1/2-wavelength shorted stub of 75-ohm cable in parallel with the 50-ohm cable where it attaches to the antenna

Here we have a 2:1 SWR problem, so we will need a matching network that splits the difference so we want a line with a 75-Ω impedance. When making this kind of match, we want the network in series between the antenna and the feed line, not in parallel. This configuration makes **Answer C** the correct choice.

E9E11 What is the primary purpose of phasing lines when used with an antenna having multiple driven elements?
- A. It ensures that each driven element operates in concert with the others to create the desired antenna pattern
- B. It prevents reflected power from traveling back down the feed line and causing harmonic radiation from the transmitter
- C. It allows single-band antennas to operate on other bands
- D. It creates a low-angle radiation pattern

When an antenna system has multiple driven elements, the elements must have the proper phase relationship to make the desired pattern. Using a phasing line gives the individual elements the proper phase relationship to make the pattern. **Answer A** captures this reasoning.

9.8   E9F - Transmission Lines

9.8.1   Overview

The *Transmission Lines* question group in Subelement E9 tests you on the physical properties of different transmission line families. The *Transmission Lines* group covers topics such as
- Characteristics of open and shorted feed lines
- Coax versus open-wire feed lines
- Velocity factor, electrical length, and coaxial cable dielectrics

The test producer will select one of the 13 questions in this group for your exam.

9.8.2   Questions

E9F01 What is the velocity factor of a transmission line?
  A. The ratio of the characteristic impedance of the line to the terminating impedance
  B. The index of shielding for coaxial cable
  C. The velocity of the wave in the transmission line multiplied by the velocity of light in a vacuum
  D. The velocity of the wave in the transmission line divided by the velocity of light in a vacuum

The velocity factor divides the transmission speed in the line by that of free space, so **Answer D** is the correct choice. Answer C is incorrect because it uses multiply instead of divide. Answers A and B are to distract you.

E9F02 Which of the following has the biggest effect on the velocity factor of a transmission line?
  A. The termination impedance
  B. The line length
  C. Dielectric materials used in the line
  D. The center conductor resistivity

As we see in Table 9.2, the transmission line's dielectrics determine the velocity factor, so **Answer C** is the correct choice. The other factors do not determine the velocity factor, so they are there to distract you.

E9F03 Why is the physical length of a coaxial cable transmission line shorter than its electrical length?
  A. Skin effect is less pronounced in the coaxial cable
  B. The characteristic impedance is higher in a parallel feed line
  C. The surge impedance is higher in a parallel feed line
  D. Electrical signals move more slowly in a coaxial cable than in air

Electrically, the only true statement is **Answer D** because the electromagnetic radiation moves slower in the transmission line than in free space.

E9F04 What impedance does a 1/2-wavelength transmission line present to a generator when the line is shorted at the far end?
  A. Very high impedance
  B. Very low impedance
  C. The same as the characteristic impedance of the line
  D. The same as the output impedance of the generator

From Table 9.3, we can see that the correct answer for this configuration is a very low impedance, so **Answer B** is the right choice.

E9F05 What is the approximate physical length of a solid polyethylene dielectric coaxial transmission line that is electrically 1/4 wavelength long at 14.1 MHz?

    A. 10.6 meters

    B. 5.3 meters

    C. 4.3 meters

    D. 3.5 meters

The 14.1-MHz frequency corresponds to a free-space wavelength of 21.3 m. Since we wish to have a ¼-wave line, the matching free-space length is 5.3 m. We multiply this by the typical velocity factor from Table 9.2, 0.66 in this case, to give 3.5 m. This result makes **Answer D** the correct choice.

E9F06 What is the approximate physical length of an air-insulated, parallel conductor transmission line that is electrically 1/2 wavelength long at 14.10 MHz?

    A. 7.0 meters

    B. 8.5 meters

    C. 10.6 meters

    D. 13.3 meters

Here we apply the same analysis as the previous question. The free-space wavelength is 21.3 m. We desire a ½-wavelength line, so the free-space length is 10.6 m. We multiply this by the typical velocity factor from Table 9.2, 0.99 in this case, to give 10.5 m. This result makes **Answer C** the correct choice.

E9F07 How does ladder line compare to small-diameter coaxial cable such as RG-58 at 50 MHz?

    A. Lower loss

    B. Higher SWR

    C. Smaller reflection coefficient

    D. Lower velocity factor

The ladder line will have a higher velocity factor than the coax, so Answer D is incorrect. The lower loss in **Answer A** is a correct statement. Answers B and C are untrue statements.

E9F08 Which of the following is a significant difference between foam dielectric coaxial cable and solid dielectric cable, assuming all other parameters are the same?

    A. Foam dielectric has lower safe operating voltage limits

    B. Foam dielectric has lower loss per unit of length

    C. Foam dielectric has higher velocity factor

    D. All of these choices are correct

Each statement in Answers A, B, and C is correct, so **Answer D** is the right choice.

E9F09 What is the approximate physical length of a foam polyethylene dielectric coaxial transmission line that is electrically 1/4 wavelength long at 7.2 MHz?
A. 10.4 meters
B. 8.3 meters
C. 6.9 meters
D. 5.2 meters

The free-space wavelength is 41.6 m. We desire a ¼-wavelength line, so the free-space length is 10.4 m. We multiply this by the typical velocity factor from Table 9.2, 0.80 in this case, to give 8.3 m. This computation makes **Answer B** the right choice. Answer C is for solid dielectric, and the others are distractions.

E9F10 What impedance does a 1/8-wavelength transmission line present to a generator when the line is shorted at the far end?
A. A capacitive reactance
B. The same as the characteristic impedance of the line
C. An inductive reactance
D. Zero

Here we have a series of questions with various line lengths and terminations. Table 9.3 provides a guide for how the lines operate under various conditions. From the table, we see that the inductive reactance of **Answer C** is correct.

E9F11 What impedance does a 1/8-wavelength transmission line present to a generator when the line is open at the far end?
A. The same as the characteristic impedance of the line
B. An inductive reactance
C. A capacitive reactance
D. Infinite

From Table 9.3, we can see that the correct answer for this configuration is capacitive, so **Answer C** is the right choice.

E9F12 What impedance does a 1/4-wavelength transmission line present to a generator when the line is open at the far end?
A. The same as the characteristic impedance of the line
B. The same as the input impedance to the generator
C. Very high impedance
D. Very low impedance

From Table 9.3, we can see that the correct answer for this configuration is a very low impedance as in **Answer D**.

E9F13 What impedance does a 1/4-wavelength transmission line present to a generator when the line is shorted at the far end?
A. Very high impedance
B. Very low impedance
C. The same as the characteristic impedance of the transmission line
D. The same as the generator output impedance

Again, from Table 9.3, we can see that the correct answer for this configuration is a very high impedance as in **Answer A**.

9.9   E9G - The Smith Chart

9.9.1   Overview

The Smith Chart question group in Subelement E9 looks at the ways users plot the characteristics of devices on the Smith chart. The test producer will select one of the 11 questions in this group for your exam.

9.9.2   Questions

E9G01 Which of the following can be calculated using a Smith chart?
A. Impedance along transmission lines
B. Radiation resistance
C. Antenna radiation pattern
D. Radio propagation

Designers use a Smith chart to plot impedances. From this, we can obtain the transmission line impedance making **Answer A** the correct choice. The other options are all good to know, but a Smith chart will not help.

E9G02 What type of coordinate system is used in a Smith chart?
A. Voltage circles and current arcs
B. Resistance circles and reactance arcs
C. Voltage lines and current chords
D. Resistance lines and reactance chords

The mnemonic for this question is "R & R": resistance and reactance, so **Answer B** is the correct choice to answer the question. The other choices are to see if you know what a Smith chart is. Be careful with Answer D because it has R & R, but the resistance is on a circle, and reactance is on an arc, not lines and chords.

E9G03 Which of the following is often determined using a Smith chart?
   A. Beam headings and radiation patterns
   B. Satellite azimuth and elevation bearings
   C. Impedance and SWR values in transmission lines
   D. Trigonometric functions

Since designers use a Smith chart to plot impedances, **Answer C** is the correct choice. Answers A and B have different types of charts to plot that information. Trigonometric functions are to make you chuckle.

E9G04 What are the two families of circles and arcs that make up a Smith chart?
   A. Resistance and voltage
   B. Reactance and voltage
   C. Resistance and reactance
   D. Voltage and impedance

Since the Smith chart is related to impedances, **Answer C** is correct because that is the one dealing with resistance and reactance. The other choices contain voltage, which is incorrect for a Smith chart.

E9G05 Which of the following is a common use for a Smith chart?
   A. Determine the length and position of an impedance matching stub
   B. Determine the impedance of a transmission line, given the physical dimensions
   C. Determine the gain of an antenna given the physical and electrical parameters
   D. Determine the loss/100 feet of a transmission line, given the velocity factor and conductor materials

Answer A is correct because designers use the Smith chart in designing impedance matching circuits. One does not find the other parameters with the Smith chart.

E9G06 On the Smith chart shown in Figure E9-3, what is the name for the large outer circle on which the reactance arcs terminate?
   A. Prime axis
   B. Reactance axis
   C. Impedance axis
   D. Polar axis

Figure E9-3

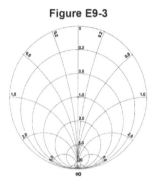

Figure 9.10: Figure E9-3.

If you noticed that the former questions always dealt with resistance and reactance, then you should be able to spot **Answer B** as the right choice because that one has reactance. Be careful with Answer C since we say that designers use a Smith chart for impedance plotting, but that is not the proper term here.

E9G07 On the Smith chart shown in Figure E9-3, what is the only straight line shown?
  A. The reactance axis
  B. The current axis
  C. The voltage axis
  D. The resistance axis

If the last question asked about the reactance axis, so this question asks about the resistance axis, and **Answer D** is the correct choice. Answer A was for the previous question, so be careful here. Designers do not plot currents and voltages directly on a Smith chart.

E9G08 What is the process of normalization with regard to a Smith chart?
  A. Reassigning resistance values with regard to the reactance axis
  B. Reassigning reactance values with regard to the resistance axis
  C. Reassigning impedance values with regard to the prime center
  D. Reassigning prime center with regard to the reactance axis

Answer C gives the correct definition because the prime center is the value designers use to normalize the other values. The actions suggested by the other answers are incorrect.

E9G09 What third family of circles is often added to a Smith chart during the process of solving problems?
  A. Standing wave ratio circles
  B. Antenna-length circles
  C. Coaxial-length circles
  D. Radiation-pattern circles

Answer A is correct since the standing wave ratio will give a measure of impedance mismatch. The Smith chart is not an antenna characteristics chart. Coaxial cables can have their reactance characteristics plotted on a Smith chart, but they do not form circles except when one winds them up.

E9G10 What do the arcs on a Smith chart represent?
A. Frequency
B. SWR
C. Points with constant resistance
D. Points with constant reactance

As we saw above, the arcs are reactance arcs, so we should be able to spot the correct choice of constant reactance in **Answer D**.

E9G11 How are the wavelength scales on a Smith chart calibrated?
A. In fractions of transmission line electrical frequency
B. In fractions of transmission line electrical wavelength
C. In fractions of antenna electrical wavelength
D. In fractions of antenna electrical frequency

One calibrates the Smith chart in terms of the electrical wavelength, so be sure to choose **Answer B** for this question.

9.10 E9H - Receiving Antennas

9.10.1 Overview

The *Receiving Antennas* question group in Subelement E9 covers long-wavelength receiving antennas and antennas for applications such as fox hunts. The *Receiving Antennas* group covers topics such as
- Radio direction-finding antennas
- Beverage antennas
- Specialized receiving antennas
- Long-wire receiving antennas

The test producer will select one of the 11 questions in this group for your exam.

9.10.2 Questions

E9H01 When constructing a Beverage antenna, which of the following factors should be included in the design to achieve good performance at the desired frequency?
A. Its overall length must not exceed 1/4 wavelength
B. It must be mounted more than 1 wavelength above ground
C. It should be configured as a four-sided loop
D. It should be one or more wavelengths long

The correct rule of thumb for the Beverage antenna is that it should be greater than one wavelength in length, so **Answer D** is the right choice.

E9H02 Which is generally true for low band (160 meter and 80 meter) receiving antennas?
  A. Atmospheric noise is so high that gain over a dipole is not important
  B. They must be erected at least 1/2 wavelength above the ground to attain good directivity
  C. Low loss coax transmission line is essential for good performance
  D. All of these choices are correct

At these wavelengths, distant thunderstorms and even radio noise from space can be high, so a significant gain will only amplify the noise too. **Answer A** is the right choice in this case.

E9H03 What is Receiving Directivity Factor (RDF)?
  A. Forward gain compared to the gain in the reverse direction
  B. Relative directivity compared to isotropic
  C. Relative directivity compared to a dipole
  D. Forward gain compared to average gain over the entire hemisphere

Directivity gives a measure of the concentration of the antenna's radiation pattern. The ratio of the forward gain to the average gain in **Answer D** is the proper ratio for the Receiving Directivity Factor (RDF). The other choices are to distract you.

E9H04 What is an advantage of placing a grounded electrostatic shield around a small loop direction-finding antenna?
  A. It adds capacitive loading, increasing the bandwidth of the antenna
  B. It eliminates unbalanced capacitive coupling to the surroundings, improving the nulls
  C. It eliminates tracking errors caused by strong out-of-band signals
  D. It increases signal strength by providing a better match to the feed line

Having well-defined nulls in a Direction Finding (DF) antenna pattern is an essential characteristic in helping to find hidden transmitters. **Answer B** captures this characteristic, so it is the correct choice.

E9H05 What is the main drawback of a small wire-loop antenna for direction finding?
  A. It has a bidirectional pattern
  B. It has no clearly defined null
  C. It is practical for use only on VHF and higher bands
  D. All these choices are correct

The bi-directional nature of the antenna, given in **Answer A**, is a valid reason for not using a wire-loop antenna; the pattern will not uniquely identify the direction. Answers B, C, and are incorrect as well.

E9H06 What is the triangulation method of direction finding?
  A. The geometric angles of sky waves from the source are used to determine its position
  B. A fixed receiving station plots three headings to the signal source
  C. Antenna headings from several different receiving locations are used to locate the signal source
  D. A fixed receiving station uses three different antennas to plot the location of the signal source

To perform triangulation, one combines headings from several diverse physical locations to isolate the transmitter location, so **Answer C** is the correct choice. Answer A would require very expensive and highly-directional antennas, generally not available to amateurs. Answer B, as stated, might not work because it has just one antenna, so it will not see multiple, intersecting paths. Answer D is not a good choice because the antennas need a wide separation to have sufficient resolution.

E9H07 Why is RF attenuation used when direction-finding?
  A. To narrow the receiver bandwidth
  B. To compensate for isotropic directivity and the antenna effect of feed lines
  C. To increase receiver sensitivity
  D. To prevent receiver overload which reduces pattern nulls

If you think about it, the receiver in the "fox hunt" will need to work with received signals over an extensive signal strength range as the hunter gets closer to the fox. Since the receiver has a limited reception signal strength range, an attenuator is convenient for extending the receiver's dynamic range by attenuating signals from a nearby fox. This approach makes **Answer D** the best choice.

E9H08 What is the function of a sense antenna?
  A. It modifies the pattern of a DF antenna array to provide a null in one direction
  B. It increases the sensitivity of a DF antenna array
  C. It allows DF antennas to receive signals at different vertical angles
  D. It provides diversity reception that cancels multipath signals

If a DF antenna can receive a signal from all directions, it is hard to isolate in which direction to concentrate. If you can produce a null in a specific direction, you can use that null to isolate the direction to the source. A sense antenna, as in **Answer A**, provides this null in the DF antenna pattern.

E9H09 What is a Pennant antenna?
- A. A four-element, high-gain vertical array invented by George Pennant
- B. A small, vertically oriented receiving antenna consisting of a triangular loop terminated in approximately 900 ohms
- C. A form of rhombic antenna terminated in a variable capacitor to provide frequency diversity
- D. A stealth antenna built to look like a flagpole

Answer A is to make you chuckle. A pennant antenna has a triangular shape, like a pennant, so **Answer B** ought to be easy to spot as the correct answer. The others are to distract you.

E9H10 How can the output voltage of a multiple turn receiving loop antenna be increased?
- A. By reducing the permeability of the loop shield
- B. By utilizing high impedance wire for the coupling loop
- C. By winding adjacent turns in opposing directions
- D. By increasing the number of turns and/or the area

The strength of the signal coming from a loop antenna is proportional to the area of the antenna, and the number of loop turns. The antenna's output voltage increases as both design parameters increase. From this property, we can see that Answers B and C work opposite to at least one measurement trend. **Answer D** has the increasing parameters both going in the right direction, so this is the correct choice. Answer A is electrically incorrect.

E9H11 What feature of a cardioid pattern antenna makes it useful for direction finding?
- A. A very sharp peak
- B. A very sharp single null
- C. Broadband response
- D. High radiation angle

An antenna with either a sharp peak or a sharp null is good for DF because the operator can use the peak or null to "point" to the RF source. Broadband responses are not suitable for direction finding because they are not directionally sensitive. The cardioid has a sharp null, so **Answer B** has the correct justification for using a cardioid pattern. Answer A is an incorrect choice because a cardioid pattern does not have a sharp peak. Answer D is incorrect because a cardioid pattern does not have a high takeoff angle.

Chapter 10

E0 — Safety

10.1 Introduction

The *Safety* subelement builds on the amateur radio safety practices we learned in the Technician and General Class studies. The *Safety* subelement has one question group that will generate one question on the Extra Class license examination.

10.2 Radio Engineering Concepts

Maximum Permissible Exposure   The Federal Communications Commission (FCC) and professional groups have developed standards for maximum permissible exposure to Radio Frequency (RF) radiation (see https://tran sition.fcc.gov/oet/info/documents/bulletins/oet65/oet65b.pdf). Table 10.1 summarizes the FCC limits. The limits are a function of operating frequency because the parts of the human body respond differently at different frequencies. The limits are based on the power density of the radiation when it encounters the human body and the *duty cycle* of the exposure. The duty cycle is the average time the radiation is "on" compared with the total measurement time span, usually 6 minutes.

In Part 97, the FCC requires that licensees understand if their station will cause an excessive RF exposure. This means that the licensee must perform a radiation evaluation when they first establish their station and whenever they make any major changes to the station, such as a change in the antenna type or placement. A licensee can perform a quick exposure estimate on the web using the ARRL RF Exposure Calculator (http://arrl.org/rf-exposure-calcula tor). Many licensees will find that the analysis shows their station does not need a detailed assessment because they operate at a power level with sufficient distance from the operator and the general public that the station environment is safe. The licensee must perform a more detailed evaluation if the station

Table 10.1: Maximum Exposure Limits for Occupational/Controlled Exposure (OET Bulletin 65, August 1997; `https://transition.fcc.gov/oet/info/do cuments/bulletins/oet65/oet65.pdf`)

Frequency (MHz)	Electric Field Strength (E) (V/m)	Magnetic Field Strength (H) (A/m)	Power Density (S) (mW/cm^2)	Averaging Time (minutes)
0.3 – 3.0	614	1.63	100	6
3.0 – 30	$1842/f$	$4.89/f$	$900/f^2$	6
30 – 300	61.4	0.163	1.0	6
300 – 1500	—	—	$f/300$	6
1500 – 100,000	—	—	5	6

exceeds the safe exposure distance. The FCC provides a four-page worksheet at the end of the OET Bulletin 65 to assist with this evaluation. You can find more information on websites such as `https://www.arrl.org/rf-exposure` and `http://www.arrl.org/fcc-rf-exposure-regulations-the-station-e valuation`.

10.3   E0A - Safety

10.3.1   Overview

The *Safety* question group in Subelement E0 tests how you implement safe practices in your radio operations. The *Safety* group covers topics such as
1. RF radiation hazards
2. Hazardous materials
3. Grounding

The test producer will select one of the 11 questions in this group for your exam.

10.3.2   Questions

E0A01 What is the primary function of an external earth connection or ground rod?
A. Reduce received noise
B. Lightning protection
C. Reduce RF current flow between pieces of equipment
D. Reduce RFI to telephones and home entertainment systems

The ground rod and Earth connection provide lightning protection making **Answer B** the correct choice. The other choices might be secondary results, but the question asks about the primary reason, so they are not the best choice to answer the question.

E0A02 When evaluating RF exposure levels from your station at a neighbor's home, what must you do?
A. Ensure signals from your station are less than the controlled Maximum Permitted Exposure (MPE) limits
B. Ensure signals from your station are less than the uncontrolled Maximum Permitted Exposure (MPE) limits
C. Ensure signals from your station are less than the controlled Maximum Permitted Emission (MPE) limits
D. Ensure signals from your station are less than the uncontrolled Maximum Permitted Emission (MPE) limits

Amateur best practice says that you should make sure that the signals from your station are lower than the uncontrolled Maximum Permissible Exposure (MPE) limits as given in **Answer B**. The uncontrolled levels are for use with the general public, like your neighbor. Be careful with Answers C and D because they have similar-sounding incorrect definitions for MPE.

E0A03 Over what range of frequencies are the FCC human body RF exposure limits most restrictive?
A. 300 kHz to 3 MHz
B. 3 to 30 MHz
C. 30 to 300 MHz
D. 300 to 3000 MHz

If we examine Table 10.1, we can see that the frequency range with the lowest permitted power density for exposure is the 30 to 300 MHz region making **Answer C** the correct choice. The other regions permit a higher power level for exposure.

E0A04 When evaluating a site with multiple transmitters operating at the same time, the operators and licensees of which transmitters are responsible for mitigating over-exposure situations?
A. Only the most powerful transmitter
B. Only commercial transmitters
C. Each transmitter that produces 5 percent or more of its MPE limit in areas where the total MPE limit is exceeded
D. Each transmitter operating with a duty-cycle greater than 50 percent

As you may suspect, good engineering practice dictates the owners of all transmitters should participate in the site evaluation. However, this question addresses the minimal requirements for responsibility in mitigating over-exposure when there are multiple active transmitters as found in OET Bulletin 65. When deciding on the inclusion of a transmitter, the 5% rule given in **Answer C** is the permitted exception level for including a transmitter in the multiple station evaluation, which makes this the choice to answer the question.

E0A05 What is one of the potential hazards of operating in the amateur radio microwave bands?
A. Microwaves are ionizing radiation
B. The high gain antennas commonly used can result in high exposure levels
C. Microwaves often travel long distances by ionospheric reflection
D. The extremely high frequency energy can damage the joints of antenna structures

Each statement made in Answers A, C, and D is untrue. Microwaves can result in high exposure levels when combined with high gain antennas, so **Answer B** is the right choice.

E0A06 Why are there separate electric (E) and magnetic (H) field MPE limits?
A. The body reacts to electromagnetic radiation from both the E and H fields
B. Ground reflections and scattering make the field impedance vary with location
C. E field and H field radiation intensity peaks can occur at different locations
D. All of these choices are correct

Each statement in Answers A, B, and C is correct, so **Answer D** is the best choice.

E0A07 How may dangerous levels of carbon monoxide from an emergency generator be detected?
A. By the odor
B. Only with a carbon monoxide detector
C. Any ordinary smoke detector can be used
D. By the yellowish appearance of the gas

The correct answer is obvious: use a carbon monoxide detector as in **Answer B**. The other options are untrue statements.

E0A08 What does SAR measure?
A. Synthetic Aperture Ratio of the human body
B. Signal Amplification Rating
C. The rate at which RF energy is absorbed by the body
D. The rate of RF energy reflected from stationary terrain

The Specific Absorption Rate (SAR) is the rate at which the body absorbs radiation, so **Answer C** is the right choice.

E0A09 Which insulating material commonly used as a thermal conductor for some types of electronic devices is extremely toxic if broken or crushed and the particles are accidentally inhaled?
A. Mica
B. Zinc oxide
C. Beryllium Oxide
D. Uranium Hexafluoride

The beryllium oxide of **Answer C** is the right choice. Mica and zinc oxide are not hazardous. Manufacturers do not use uranium hexafluoride in this context.

E0A10 What toxic material may be present in some electronic components such as high voltage capacitors and transformers?
A. Polychlorinated biphenyls
B. Polyethylene
C. Polytetrafluroethylene
D. Polymorphic silicon

This question is referring to the Polychlorinated Biphenyl (PCB) materials that the manufacturer may have used, so **Answer A** is the right choice.

E0A11 Which of the following injuries can result from using high-power UHF or microwave transmitters?
A. Hearing loss caused by high voltage corona discharge
B. Blood clotting from the intense magnetic field
C. Localized heating of the body from RF exposure in excess of the MPE limits
D. Ingestion of ozone gas from the cooling system

The RF radiation may cause localized heating making **Answer C** the right choice. The other options are distractions.

Appendix A

Acronyms, Abbreviations, and Symbols

A.1 Radio Acronyms and Abbreviations

AC   Alternating Current

ADC   Analog-to-Digital Converter

AF   Audio Frequency

AFSK   Audio Frequency Shift Keying

ALE   Automatic Link Enable

AM   Amplitude Modulation

AMTOR   Amateur Teleprinting Over Radio

APRS   Automatic Packet Reporting System

ARQ   Automatic Repeat reQuest

BDR   Blocking Dynamic Range

BPF   Band Pass Filter

BPSK   Binary Phase Shift Keying

BJT   Bipolar Junction Transistor

CEPT   European Conference of Postal and Telecommunications Administrations

CFR   Code of Federal Regulations

CMOS	Complementary Metal Oxide Semiconductor
CW	Continuous Wave
DAC	Digital-to-Analog Converter
DC	Direct Current
DF	Direction Finding
DIP	Dual In-line Package
DRM	Digital Radio Mondial
DSB	Dual Sideband
DSB-SC	Dual Sideband - Suppressed Carrier
DSP	Digital Signal Processor
DSSS	Direct Sequence Spread Spectrum
DTMF	Dual Tone Multifrequency
DX	Distant
EHF	Extremely High Frequency
EIRP	Effective Isotropic Radiated Power
ELF	Extremely Low Frequency
EME	Earth-Moon-Earth
ERP	Effective Radiated Power relative to a dipole
FBW	Fractional Band Width
FCC	Federal Communications Commission
FDM	Frequency Division Multiplexing
FEC	Forward Error Correction
FET	Field Effect Transistor
FHSS	Frequency Hopping Spread Spectrum
FIR	Finite Impulse Response
FM	Frequency Modulation
FRN	FCC Registration Number
FSK	Frequency Shift Keying

GaAs	Gallium Arsenide
GFCI	Ground Fault Circuit Interrupter
GPS	Global Positioning System
HEO	High Earth Orbit
HF	High Frequency
HPBW	Half Power Beam Width
HPF	High Pass Filter
HST	Hubble Space Telescope
IARP	International Amateur Radio Permit
IF	Intermediate Frequency
IMD	Intermodulation Distortion
ITU	International Telecommunication Union
LED	Light Emitting Diode
LEO	Low Earth Orbit
LF	Low Frequency
LPF	Low Pass Filter
LSB	Lower Side Band
MDS	Minimum Discernible Signal
MF	Medium Frequency
MFSK	Multiple Frequency Shifty Keying
MMIC	Monolithic Microwave Integrated Circuit
MOSFET	Metal Oxide Semiconductor Field Effect Transistor
MPE	Maximum Permissible Exposure
MUF	Maximum Usable Frequency
NCVEC	National Conference of Volunteer Examiner Coordinators
NTSC	National Television System Committee
OCFD	Off Center Fed Dipole
OFDM	Orthogonal Frequency Division Multiplexing

PCB	Polychlorinated Biphenyl
PEP	Peak Envelope Power
PF	Power Factor
PIN	Positive-Intrinsic-Negative
PLD	Programmable Logic Device
PLL	Phase Locked Loop
PLCC	Plastic-Leaded Chip Carrier
PM	Phase Modulation
PSK	Phase Shift Keying
QPSK	Quadrature Phase Shift Keying
RDF	Receiving Directivity Factor
RF	Radio Frequency
RMS	Root Mean Square
RR	Radio Regulations
RTTY	Radio TeleType
SAR	Specific Absorption Rate
SDR	Software Defined Radio
SHF	Super High Frequency
SNR	Signal-to-Noise Ratio
SOT	Small Outline Transistor
SS	Spread Spectrum
SSB	Single Sideband
SSTV	Slow-Scan Television
STA	Special Temporary Authority
SWR	Standing Wave Ratio
TDM	Time Division Multiplexing
TTL	Transistor-Transistor Logic
UHF	Ultra High Frequency

UI	Unnumbered Information
US	United States
USB	Upper Side Band
UV	Ultra Violet
VCO	Voltage Controlled Oscillator
VE	Volunteer Examiner
VEC	Volunteer Examiner Coordinator
VFO	Variable Frequency Oscillator
VHF	Very High Frequency
VIS	Vertical Interval Signaling
VLF	Very Low Frequency
VOCAP	Voice of America Coverage Analysis Program
VSB	Vestigial Side Band

A.2 Functions, Symbols, Units, and Variables

A	ampere
B	Transmitted Bandwidth, Hz
B	Magnetic flux density, T
B	Susceptance, S
β	Modulation Index
bps	Bits per second
c	Speed of Light, $299\,792\,458\,\mathrm{m/s}$
C	coulomb
C	Capacitance, F
cm	centimeter
D	Deviation Ratio
Δ	Frequency Shift, Hz
Δf	Frequency Deviation, Hz

dB	decibel
dBW	decibel Watt
dBm	decibel milliWatt
f	Frequency, Hz
f_{IMD}	Intermod Frequency, Hz
f_m	Tone Frequency, Hz
f_r	Resonant Frequency, Hz
F	farad
G	Gain, unitless
G	Conductance, S
Gbps	1 000 000 000 bps
GHz	1 000 000 000 Hz
H	henry
Hz	hertz
i	Current, A
k	Boltzmann's constant, $1.380\,648\,52 \times 10^{-23}$ W/(K − Hz) or -228.6 dBW/(K − Hz)
K	Absolute temperature scale
kbps	1000 bps
kHz	1000 Hz
km	1000 m
L	Inductance, H
λ	Wavelength, m
m	meter
$m(t)$	Message signal
M	Message Signal Bandwidth, Hz
mA	0.001 A
μA	0.000 001 A

Mbps	1 000 000 bps
MHz	1 000 000 Hz
mH	0.001 H
mV	0.001 V
mW	0.001 W
μF	0.000 001 F
μH	0.000 001 H
μV	0.000 001 V
μW	0.000 001 W
N	Noise Power, W
N_o	Noise density, W/Hz
Q	Quality Factor
θ	Phase angle
T_{sys}	Noise Temperature, K
P	Power, W
pF	0.000 000 000 001 F
R	Resistance, Ω
R_b	Bit Rate, bps
RC	Resistor-Capacitor
RLC	Resistor-Inductor-Capacitor
s	second
S	Siemens
sps	Symbols per second
$s(t)$	Carrier signal
T	tesla
τ	Time constant, s
V	volt
Ω	ohm

W	watt
X	Reactance, Ω
Y	Admittance, S
Z	Impedance, Ω

Made in United States
Troutdale, OR
12/04/2023